Utilization of Microorganisms in Meat Processing

INNOVATION IN MICROBIOLOGY SERIES

Series Editor: **Dr A. N. Sharpe,** *Health and Welfare, Ottawa, Canada*

1. The Direct Epifluorescent Filter Technique for the rapid enumeration of micro-organisms
   G. L. Pettipher

2. Utilization of Microorganisms in Meat Processing:
   A handbook for meat plant operators
   Jim Bacus

Utilization of Microorganisms in Meat Processing

A handbook for meat plant operators

Jim Bacus, Ph.D.

American Bacteriological and
Chemical Research Corporation
Gainesville, Florida, United States

RESEARCH STUDIES PRESS LTD.
Letchworth, Hertfordshire, England
JOHN WILEY & SONS INC.
New York · Brisbane · Chichester · Toronto · Singapore

RESEARCH STUDIES PRESS LTD.
58B Station Road, Letchworth, Herts. SG6 3BE, England

Marketing and Distribution:

Australia, New Zealand, South-east Asia:
Jacaranda-Wiley Ltd., Jacaranda Press
JOHN WILEY & SONS INC.
GPO Box 859, Brisbane, Queensland 4001, Australia

Canada:
JOHN WILEY & SONS CANADA LIMITED
22 Worcester Road, Rexdale, Ontario, Canada

Europe, Africa:
JOHN WILEY & SONS LIMITED
Baffins Lane, Chichester, West Sussex, England

North and South America and the rest of the world:
JOHN WILEY & SONS INC.
605 Third Avenue, New York, NY 10158, USA

Library of Congress Cataloguing in Publication Data:

Bacus, Jim.
Utilization of microorganisms in meat processing.
(Innovation in microbiology series; 2)
Bibliography: p.
Includes index.
1. Meat industry and trade. 2. Micro-organisms—Industrial applications. I. Title. II. Series.
TS1962.B28 1984 664'.924 83-17834
ISBN 0 471 90312 4 (Wiley)

British Library Cataloguing in Publication Data:

Bacus, Jim
Utilization of microorganisms in meat processing
1. Meat industry and trade
I. Title
664'.9 TS1960
ISBN 0 86380 011 4

ISBN 0 86380 011 4 (Research Studies Press Ltd.)
ISBN 0 471 90312 4 (John Wiley & Sons Inc.)

Printed in Great Britain

Table of Contents

Preface

Microorganisms generally have been regarded as the "enemy" in the processing of meat. After the live animal has been sacrificed, it is a race between man and microorganism as to whom will consume the meat tissue. Although fermented meat products, where beneficial microorganisms actually enhance and preserve the product, have been manufactured for centuries, the respective manufacturers were unaware of the technical aspects of the process. As early as 1921, researchers recognized the contribution of microorganisms to meat processing. Later studies in the 1940's and 1950's further documented the role of microorganisms, and these led to the suggested usage of pure microbial cultures to achieve greater consistency. Although commercial meat cultures have been available to processors since 1960, only within the last 4 to 5 years have many processing establishments converted to the use of pure cultures. Fermented meat research is difficult due to the nature of the raw materials and, although the literature on the subject is extensive, it has not been applied or even adequately explained to the sausage processor.

This book attempts to merge the basic knowledge of microorganisms and meat fermentation with the practical aspects and observations in meat processing. The terminology and subject matter reflects the desire for the book to serve as a reference for both the meat processor and the meat scientist.

Jim Bacus

Foreword

Amongst our options when we attempt to minimize the deaths, illness and economic losses caused by microbial contamination in foods are what might be regarded as two diametrically opposed avenues. There is, as it were, a bludgeon approach in which we see the development of manufacturing, distribution and retailing practices designed to keep microbial contamination of all kinds to a minimum before the food is consumed. It is the commonest approach, and most of our microbiological standards, manufacturing procedures and public health regulations are directed towards this end. And it is certainly logical that minimizing the numbers of spoilage or pathogenic microorganisms in manufactured foods will, on average, minimize the number of consumers at risk or the quantity of food lost through spoilage. Enormous development effort has gone into, and continues to take place, in this area of food science and technology.

The logical goal in this approach is the complete exclusion of microorganisms from foods before consumption. One cannot help feeling, however, that attempting to approach closer and closer to sterility in manufactured foods is a little like trying to approach the absolute zero in temperature--the closer one approaches, the harder becomes the task of going further--or even staying still. A great disadvantage of this approach is the fact that as food quality increases, the extent of the normal microbial interactions decreases. A few pathogenic microorganisms inadvertently allowed in the food may thus find themselves in a growth environment completely without competition, into which they can multiply unchecked with obvious consequences to the consumer.

There is a second, and subtler approach, one handed down for centuries by certain peoples, exploiting the ability of various harmless members of the Lactobacillaceae, Micrococcaceae, etc., to inhibit the grwoth of less desirable organisms. The inhibition may be caused by the altering of the environment of the food (pH, E_h, etc.) into one unfavorable for these other organisms, to the formation of specific antibacterial substances such as the bacteriocins secreted by some lactobacilli, or to more complex forms of antagonism. Regardless of the mechanism, this approach requires the deliberate addition of microorganisms to a food to bring about its safety and stability.

On the one hand the starter culture may be allowed to grow to completion in the product, in which case the metabolic consequences of the microorganisms have an overt effect on the product. In fact, they sometime *are* the product. The traditional antecedent of the starter culture was recycling a part of one finished product into the raw materials of the next batch. It was successful much of the time but, because the culture was ill-defined, disastrous on occasion. The complex mixtures of organisms used in the natural approach may lead to unique and characteristic flavors or other properties not yet completely matched by the purified starter cultures of today--nevertheless, the uniformity and reliability of modern starters leads to a great increase in overall safety.

On the other hand there is the possibility of adding a starter culture at low level to the product, without allowing it extensive growth. The added culture thus has a minimal effect on the organoleptic properties of the product, until such time as an abuse occurs, perhaps through storage at an incorrect temperature. In the case of bacon, for example, a normally unobtrusive starter might, in the event of improper storage of low-nitrite product, insure that the pH shifts below a level at which botulinal toxins may form. As far as the consumer is concerned the product may spoil, but at least it has failed in a safe manner.

It seems to me that, notwithstanding the quantity of basic research carried out on top of centuries of traditional processes, we are just in the infancy of a new technology whereby foods are preserved or rendered fail-safe by the positive use of microorganisms against microorganisms. It is surely more satisfying to work *with* microbes this way, than taking the brute force approach of trying to exclude them increasingly from the product. It also makes economic sense as the energy costs of traditional preservation processes makes them less attractive.

Through his considerable involvement and experience with the development and use of starter cultures for meat processing, Dr. Bacus is in an excellent position to write authoritatively about the science. His book gives us all sides of the story--not only a fine overview of the positive uses of microorganisms in meat preservation and the development work in progress, but an extensive "cookbook" (literally) for microbiologists and technologists engaged in the manufacture of fermented meat products. In so doing, he provides a valuable workbase for this burgeoning area of food science.

A.N. Sharpe

List of Tables

List of Illustrations

CHAPTER 1
Historical Perspective

Fermented Foods

Fermentation and drying are the oldest methods of food preservation known to mankind. Although unaware of the process, man has relied on fermentation for centuries as an effective and economical way to preserve foods. The process still remains one of our most important methods, although most people are still unaware that many of the products they consume are prepared and/or preserved by the fermentation process.

A significant advance in man's history was the transition from food gathering to food production. Man learned that the proper handling and storage of many perishable food stuffs brought about changes in their physical, chemical and organoleptic characteristics that proved desirable, and yielded greater product stability. Changes in milk during storage could result in a smooth, desirable acid curd which proved edible and wholesome. Certain fruit juices yielded very acceptable and exhilarating beverages. Many vegetable materials, when properly packed and stored, resulted in a wide variety of stable food products, and dried seeds could be ground, blended with salt and water and subsequently heated to improve flavor and quality. Meats could also be ground, mixed with salt and spices and held at cool temperatures to provide a wide variety of sausage products that were both safe for consumption and were acceptable to the palate.

Today we realize that these products undergo a biological fermentation process whereby specific bacteria and/or yeasts, transform sugars to a variety of acids and/or alcohols that inhibit undesirable microorganisms, including food pathogens. These fermented foods owe their production, flavor, texture, nutritional, stability and/or other characteristics to the activities of these beneficial microorganisms. Historically, these changes in the food would occur through the accidental introduction of the microflora from the environment. Man learned that through control of the environmental handling and storage conditions of the food, he could generally bring about the desirable changes. However, lack of control would result in undesirable and/or unsafe product.

Starter Cultures

In recent years, the desirable microorganisms for many of these fermented products have been isolated, identified, propagated separately in the laboratory, and subsequently reintroduced as starter cultures. These single or multiple strain starter cultures are added to achieve a controlled fermentation that ensures and enhances the production of the desired end products. Samples of the respective microorganisms and the fermented foods are extensive (Tables 1-4).

The utilization of microorganisms as starter cultures in the manufacture of various fermented dairy, cereal, and beverage products has been well documented (Smith and Palumbo, 1981). The respective food industries have successfully employed commercial starter cultures for many years. Although it has been customary to think of starter cultures primarily as agents of acid, alcohol, and/or carbon dioxide production, it is becoming more apparent that these end products may not be the only changes produced by these microorganisms in the respective foodstuffs. The nutritional content can be altered, distinctive flavors are produced, microbial by-products are formed that are antagonistic to undesirable microorganisms, and the starters may contain components that are beneficial to the consumer.

The fermented vegetable and meat industries, have been much slower to adopt the use of pure cultures of specific microorganisms. Until recent years, most manufacturers of fermented meat products in the United States did not utilize commercially available meat cultures. This is still true in most other countries. A partial explanation lies in the fact that research in meat technology, particularly fermented sausages, has been left behind research in many other areas of food processing, i.e. milk technology. Meat as a research material is much more difficult to examine than milk or dairy products, because of its nonuniformity. Investigations into the effects of microbial pure cultures also present difficulties since it is not possible to experiment with sterile or even pasteurized material, as can be done with dairy products. The effects of random microbial contaminants in meat products cannot be eliminated. Raw meats routinely may contain several millions per gram of these undesirable microorganisms and their influence on chemical changes cannot be overlooked.

Meat Processing

The origin of meat processing probably occurred when primitive man first realized that he must either rapidly consume the fresh meat after slaughter, or it would spoil, and be unfit for later consumption. This observation led him to either cooking the meat to prolong the keeping quality and/or other forms of further processing. The ancient Egyptians recorded the preservation of meat by salting and sundrying. The early Romans are credited with first using ice and snow to preserve food. The preparation of sausage by cutting or grinding the meat, seasoning it with salt and spices, and drying it in rolls

Table 1. Microorganisms used as additives in dairy products (from Smith and Palumbo, 1981).

Product	Microorganisms Added
Cheese	
Parmesan, Romano	mixture of *Lactobacillus bulgaricus* and *Streptococcus thermophilus*
Cheddar, Colby	*S. lactis*; *S. cremoris*
Swiss, Emmenthaler	mixture of *L. bulgaricus* (or *L. lactis* or *L. helveticus*) and *S. thermophilus* and *Propionibacterium shermanii*
Provolone	mixture of heat-resistant *Lactobacillus* species and *S. thermophilus*
Blue, Gorgonzola, Roquefort, Stilton	*S. lactis* plus *Penicillium roqueforti*
Camembert	*lactic streptococci* plus *Penicillium camemberti*
Brick, Limburger	mixture of *S. thermophilus* and *S. cremoris*; mixture of *S. thermophilus* and *L. bulgaricus*
Brick, Limburger	mixture of *S. lactis* and *S. thermophilus*
Muenster	mixture of *S. thermophilus* and *Lactobacillus* species
Gouda, Edam	*S. lactis*
Mozzarella	mixture of heat-resistant *Lactobacillus* species and *S. thermophilus*
Cottage cheese, cream cheese	*S. lactis* or *S. cremoris*; mixture of *S. lactis* or *S. cremoris* and *S. diacetilactis* (or *Leuconostoc* species)
Fermented milks	
Bulgarian buttermilk	*L. bulgaricus*
Acidophilus milk	*L. acidophilus*
Buttermilk, sour cream	*S. lactis*; mixture of *S. cremoris* and *Leuconostoc citrovorum* (or *L. dextranicum*)
Yogurt	mixture of *L. bulgaricus* and *S. thermophilus*
Milk	
Fresh milk	*S. diacetilactis*
Fresh milk	mixture of *S. diacetilactis* and *Leuconostoc cremoris*
Fresh milk	*L. acidophilus*
10% non-fat milk solids	*L. bulgaricus*
25 and 40% non-fat milk solids	*L. acidophilus*
Miscellaneous	
Butter	mixture of *S. diacetilactis* and *S. lactis*
Cream dressing for cottage cheese	*S. lactis*; *S. cremoris*; *S. diacetilactis*

Table 2. Microorganisms used as food additives in vegetable and fruit products (from Smith and Palumbo, 1981).

Product (vegetable or fruit)	Microorganisms
Pickles	
Carrots	mixture of *Lactobacillus plantarum*, *L. brevis*, *Leuconostoc mesenteroides*, and *Pediococcus cerevisiae*
Cucumbers	mixture of *P. cerevisiae*, *L. plantarum* and *L. brevis*; mixture of *P. cerevisiae* and *L. plantarum*; *L. plantarum*
Cucumbers-sliced	*L. plantarum*
Cucumbers-diced	*L. plantarum*
Mixed vegetables; green tomatoes; hot cherry peppers	mixture of *P. cerevisiae* and *L. plantarum*; *L. plantarum*
Various vegetables-diced	*L. plantarum*
Olives	*L. plantarum*
Sauerkraut (cabbage)	*L. plantarum*
Gari (cassava)	*L. plantarum*; mixture of *L. plantarum* and *Streptococcus* species; mixture of *L. plantarum* and *L. acidophilus*
Banana pulp	*L. bulgaricus*; *S. thermophilus*; *S. faecalis*; *L. fermentum*; *Leuconostoc mesenteroides*
Wines (various fruits; alcoholic fermentation)	*Saccharomyces cerevisiae* var. *ellipsoideus*; *Saccharomyces* species
Wines (grape; deacidification)	*Leuconostoc gracile* (*L. oenos*); *Lactobacillus hilgardii*; *Schizosaccharomyces pombe*

became an effective means to preserve the fresh meat. Sausage is one of the oldest forms of processed foods and was consumed by the ancient Babylonians, Romans, and Greeks during their military campaigns. Preserved sausages, as a meat supply, were credited as one of the main factors in the success of Caesar's legions (Pederson, 1979).

The drying of meat became very common along the shores of the Mediterranean, and many have postulated that the city of Salamis, on the east coast of Cypress, is where the term "salami" originated. Many of our present-day dry sausages owe their name to cities or regions in the area (i.e. Genoa salami, Toscano, Milano). Sausage is derived from the latin word salsus, meaning salt. The early Roman butchers cut beef and pork into small pieces, added salt and spices, stuffed them into skins or washed animal intestines, and placed them in special rooms to dry. They rapidly observed that proper control of the formulation parameters and the processing rooms resulted in better and more uniform sausages. Preparation and spicing of various sausages became a culinary art in these Mediterranean countries and later in upper Europe. These meat processing operations grew rapidly and have led to the development of our current semi-dry and dry sausages. The manufacturing practices still are considered to be more of an art, than a science.

The dry and semi-dry sausages were developed to maintain stability under the prevailing conditions of each area. The basis for processing the meat was preservation by inhibiting or deterring microbial decomposition. Artificial refrigeration was not yet available as a means of food preservation. Although unknown until more recent times, the inherent stability of these sausage products was primarily dependent on a natural fermentation whereby the meat pH was lowered, preventing the growth of undesirable microorganisms. The addition of salt and spices, and the subsequent holding and drying the meat under reduced conditions (i.e. casings) favored the growth of the desirable microorganisms that effected the necessary fermentation. It can be assumed that the ancient meat processing equipment was impregnated with these lactic microorganisms and micrococci, and served as the source inoculum.

Today, the fermented dry and semi-dry sausages represent a relatively small portion of the sausage consumed in the United States. Total sausage manufactured in United States Department of Agriculture (USDA) Establishments in 1981 was approximately 5.04 billion pounds, with dry and semi-dry sausages accounting for 249 million (4.9%) and 95 million pounds (1.9%), respectively (AMI, 1982a).

Generally, five types of sausage are recognized (Table 5). Fresh sausage, as implied, represents ground or chopped meat that is mixed with salt and spices, and is cooked immediately prior to consumption. This sausage accounts for 1.09 billion pounds, or 21.6% of the sausage manufactured in USDA Establishments. Cooked sausage is heat-processed immediately after preparation and is generally sold "ready to eat." This group is the largest category at approximately 3.2 billion

Table 3. Microorganisms used as food additives in plant seed products (from Smith and Palumbo, 1981).

Product	Cereal grain	Microorganisms added
Beer	barley (malt); corn, rice, wheat barley, sorghum grain, soybean (malt adjunct)	*Saccharomyces carlsbergensis*; *S. cerevisiae*
Saké	rice	*Aspergillus oryzae* followed by *S. saké*
Tapé ketan	rice	mixture of *Amylomyces rouxii* and *Endomycopsis burtonii*
Lao-chao	rice	mixture of *Rhizopus chinensis* and *Endomycopsis* species
Ang-kak (red rice)	rice	*Monascus purpureus*
Fermented peanut milk	water extract of peanuts	a number of individual *Lactobacillus* species; *Streptococcus thermophilus*; *Pediococcus cerevisiae*
Fermented peanut flour	peanut flour	*Actinomucor elegans*; *Aspergillus oryzae*; *Mucor hiemalis*; *Neurospora sitophila*; *Rhizopus oligosporus*
Fermented peanuts	whole peanuts	*Neurospora sitophila*; *Aspergillus oryzae*; *A. niger*; *Rhizopus oligosporus*; *R. delemar*
Ontjom	peanut press cake	Neurospora intermedia
Fermented soy milk	water extract of soybeans	a number of *Lacrobacillus* species; *S. thermophilus*; *Leuconostoc mesenteroides*
Sofu (soybean)	soybeans (Ca^{++} precipitated solids from soy milk)	*Actinomucor elegans*; *Mucor* species
	soybeans (soy milk)	*S. thermophilus*
	soybeans (soy milk)	mixture of *S. thermophilus* and *Rhizopus oligosporus*; mixture of *S. thermophilus* and *Penicillium camemberti*
Natto	soybeans	*Bacillus subtilis*
Tempeh	soybeans; wheat; residue of soybeans after making soy milk or tofu; full-fat dehulled soybean grits	*Rhizopus oligosporus*
	wheat; oats; rye; barley; rice; mixture of rice and soybeans; mixture of wheat and soybeans	*R. oligosporus*
		Neurospora species

Table 3--continued

Product	Cereal grain	Microorganisms added
Shoyu (soy sauce)	soybeans and wheat	Aspergillus oryzae followed by a mixture of Saccharomyces rouxii and Lactobacillus delbrueckii
Miso (Bean paste)	mixture of rice and soybean grits	A. oryzae followed by S. rouxii
Fermented soy protein	isolated soy protein	R. oligosporus
Mahewu (magou)	corn (maize)	Lactobacillus delbrueckii
Fermented corn meal	corn (germinated maize)	S. cerevisiae; Candida tropicalis
Ogi	corn flour (maize)	mixture of Lactobacillus plantarum, Streptococcus lactis and Saccharomyces rouxii
Bread, doughnuts, pretzels, grape nuts, rolls	wheat flour	S. cerevisiae
San Francisco sour dough French bread	wheat flour	mixture of Saccaromyces exiguus (Torulopsis holmii) and L. sanfrancisco
Soda crackers	wheat flour	mixture of S. cerevisiae and various Lactobacillus species

Table 4. Microorganisms used as food additives in meat products (adapted from Smith and Palumbo, 1981).

Product	Microorganisms added
Semi-dry fermented sausages	
Lebanon bologna	mixture of *Pediococcus cerevisiae* and *Lactobacillus plantarum*
Summer sausage	*P. cerevisiae*
Summer sausage	*P. cerevisiae* and *L. plantarum* mixture
Cervelat	*P. cerevisiae*
Cervelat	mixture of *P. cerevisiae* and *L. plantarum*
Thuringer	*P. cerevisiae*
Teewurst	*Lactobacillus* species
Pork roll	*P. cerevisiae*
Dry fermented sausages	
Pepperoni	mixture of *P. cerevisiae* and *L. plantarum*
Dry sausage	*P. cerevisiae*
European dry sausage	*Micrococcus* species
European dry sausage	mixture of *Micrococcus* species and *Lactobacillus*
Salami	mixture of *Micrococcus* species and *Lactobacillus* species
Salami	*L. plantarum*
Hard salami; genoa	*Micrococcus* species; mixture of *Micrococcus* species and *P. cerevisiae*; mixture of *Micrococcus* species and *L. plantarum*
Mold ripened salami	*Penicillium* species; *P. janthinellum*; *P. simplicissimum*; *P. cyclopium*; *P. viridicatum*
Fermented sausages	
Hot bar sausage	*P. cerevisiae*
Semi-dry turkey sausage	*Pediococcus cerevisiae*
Dry turkey sausage	*P. cerevisiae*
Dry turkey sausage	mixture of *P. cerevisiae* and *Lactobacillus plantarum*

pounds, or 63.7%. Cooked specialties and non-specific loaves amount to slightly less than 381 million pounds, or 7.6%. Fresh, smoked sausage is generally cured, uncooked sausage and accounts for only 13 million pounds, or the remaining 0.26%.

Many of these fresh and cooked sausage types have evolved in the United States with the advent of widespread refrigeration and vacuum packaging as a means of preservation. However, the dried and/or fermented sausages are still extremely popular and represent a growing market in the United States. They also account for a larger percentage of the sausage consumed in Europe and other parts of the world where more traditional products dominate the market and widespread refrigeration, distribution, and packaging are not as common.

In Europe, there are various ways to classify the many types of sausages. Germany, more than any other country, is associated with sausage manufacture and the German classification is based on the temperature treatment of either the final product or the raw materials (Table 6). Fermented sausages, generally, are non-heat treated (Table 7), although some varieties of Mortadella and cooked salami (i.e. heat treated final product) are fermented.

Meat Microflora

The healthy animal possesses effective defense mechanisms to prevent microbial invasion and subsequent growth in its tissue. These mechanisms of natural, or acquired immunity, generally destroy any invading microorganisms. As a result, normal, inner tissue is considered sterile. A few microorganisms are periodically able to combat these defense barriers and therefore, are considered pathogens of the living animal.

The slaughter of the live animal breaks down these inherent defense mechanisms, rendering the tissue very susceptible to microbial invasion and growth. Meat provides an excellent growth medium for microorganisms, and the slaughtering process provides the opportunity for extensive microbial contamination. Bleeding the animal can introduce contamination into the circulatory system, while skinning and evisceration can introduce various microbes to the exposed surfaces. The main sources of microorganisms can be from the exterior of the animal (hides, hoofs, hair) which is embedded with soil, water, feed and manure, and the intestinal tract which contributes many enteric bacteria. The processing equipment and personnel serve as intermediate carriers to transfer contamination.

The wide variety of sources yield a multitude of microbial contamination. These microorganisms have been of concern to those processing and eating meat since the beginning of recorded history. Man has constantly battled the microbe to keep it from utilizing and spoiling the meat before he can consume it.

Table 5. Sausage classification (United States).

Classification	Characteristics	Examples
Fresh sausage	Fresh meats (chiefly pork); uncured, comminuted, seasoned, and usually stuffed into casings; must be cooked fully before serving.	Fresh pork sausage Bratwurst Bockwurst
Dry and semidry sausages	Cured meats; air dried, may be smoked before drying; served cold.	Genoa salami Pepperoni
Cooked sausages	Cured or uncured meats; comminuted, seasoned, stuffed into casings, cooked and sometimes smoked; usually served cold.	Frankfurters Liver sausage Braunschweiger Bologna
Fresh, smoked sausages	Fresh meats, cured or uncured, stuffed, smoked, but not cooked; must be fully cooked before serving.	Smoked, country-style pork sausage Mettwurst Kielbasa
Cooked meat specialties	Specially prepared meat products; cured or uncured meats, cooked but rarely smoked, often made in loaves, but generally sold in sliced, packaged form; usually served cold.	Loaves Head cheese Scrapple

Table 6. Classification (German) of sausage/comminuted meat products according to heat treatment by the producer (adapted from Schut, 1978).

Group	Heat treatment		Examples
	ingredients	final product	
A Raw sausage (Rohwurst)	-	-	Bratwurst, salami Mettwurst, Zervelat
B1 Cooked sausages (Bruhwurst)	-	+	Frankfurter, Bockwurst, Bierwurst
B2 Cooked sausages (Kochwurst)	+	+	Liver sausages, Bloodsausage
B3 Jellied products (Sulze)	+	+	Kopfsülze, Hausmachersülze
C Minced meat products (Hackfleisch)	- +	- +	meat balls, hamburger

Table 7. Non-heat treated sausages (from Schut, 1978).

A 1	Raw sausage	Rohwurst
A 1_1 non-fermented	Fresh sausage	Bratwurst
	pork and beef	Thuringer (fein) (-)[1]
		Nurnberger (-) (grob)
A 1_2 fermented	-	Frischwurst
		Mettwurst
		Braunschweiger (+)
		Thuringer (+)
		Zervelat (+)
		Theewurst (+)
		Regensburger knack (+)
		Thuringer knack (+)
		Frankfurter (+)
A 11 fermented	Dry sausage	Dauerwurst
	salami types	Salami
		Italian (+) (-)
		Hungarian (+) (-)
		Zerevelat (+)
		Plockwurst (+)

+ = smoked
- = not smoked

Prompt chilling is one way to inhibit the growth of most contaminating microorganisms, and this refrigeration technique is the most widely used form of meat preservation. A selective environment is created through chilling, which allows the growth of a relatively few types of bacteria on the meat surface. This microflora is predominantly gram-negative and includes members of the genera Pseudomonas and Achromobacter. Various yeasts and mold are periodically encountered, especially in the aging of beef. Excessive growth of these psychrotrophic microorganisms results in surface slime, discoloration and off-odors in the meat. This occurs primarily at the surface since these are aerobic microorganisms, but the end-products of the metabolism can penetrate into the interior tissue. Interior spoilage is not common if the meat has been properly chilled. This type of spoilage (i.e. joint sours) is generally caused by anaerobic spore-forming bacteria, as well as a variety of gram-positive, facultative bacteria that are unable to grow at low temperatures.

A number of pathogenic microorganisms can infect the animal and be transferred to the consumer. Some of these can be readily detected postmortem (i.e. tuberculosis organisms) while others are more difficult to detect, such as Brucella and Salmonella. Many pathogens may be introduced to fresh meats during handling. Most of these fresh meat pathogens do not readily grow at proper refrigeration temperatures (i.e. below 40F) and are readily killed by ordinary cooking temperatures. If properly handled and stored, the meat tissue primarily serves as an intermediate carrier for these pathogens.

Drying, salting, curing, and other forms of further processing have been used for centuries as a means to preserve meat tissue. The ancient "Kosher" salting or sanitary method of animal slaughter and dressing was developed by the Hebrews to preserve the meat. The American Indians employed salting and drying methods to make venison jerky or buffalo pemmican that also extended the meat keeping quality. The microbiology of salted, cured meats is entirely different from that of fresh meat. The curing salts (sodium chloride, sodium nitrate and/or sodium nitrite) and subsequent handling methods create a different microenvironment in the meat that favors the growth of specific, facultative, gram-positive bacteria while inhibiting the growth of the fresh meat spoilage microorganisms that are primarily aerobic and gram-negative. This "microbial inversion," occurring in the meat during the curing process, can serve to extend the shelf-life by inhibiting the majority of the fresh meat microflora. The selection process favors the type of microorganisms that are generally present in low numbers in the fresh meat, and this extends the time required for these microbes to attain sufficient numbers that may result in other types of meat spoilage. The vacuum packaging of fresh meat to extend the shelf-life also depends on this same phenomenon, whereby the normal aerobic spoilage microorganisms are eliminated or inhibited, due to lack of available oxygen.

The cured meat microflora includes members of the genera Micrococcus, Lactobacillus, Streptococcus, Leuconostoc and Microbacterium, as well as yeasts and molds. Common types of spoilage that can occur in cured meats are as follows.

1. Surface sliming resulting from the growth of copious quantities of microorganisms on the surface and containing mixtures of the bacterial genera listed above.

2. Souring from the extensive growth and acid production of lactic acid bacteria.

3. Gassiness caused by heterofermentative lactic acid bacteria (Lactobacilli, Leuconostoc) and some yeasts that produce gas from sugar fermentation.

4. Greening of the meat pigment resulting from chemical oxidation or the production of hydrogen peroxide from various lactic acid bacteria (especially L. viridescens).

Interior discolorations of microbial origin also occur from under processing of cured meats that fails to kill the responsible microorganisms.

Many food pathogens fail to grow in cured meats since the salt and nitrite are effective inhibitors. In addition, the selective environment that favors the lactic acid bacteria allows these microorganisms to rapidly out-grow any potential competitors. This growth is usually accompanied by acid production which lowers the product pH, further inhibiting food pathogens. However, food poisoning Staphylococcus strains can compete successfully in cured meats prior to any pH reduction. Ham products have been associated with many outbreaks of staphylococcus food poisoning and, invariably, the outbreaks result from the contamination and mishandling of the ham after it has been cooked. The cooking process eliminates the normal lactic microflora and can provide food pathogens an environment relatively free of microbial competition if the food is contaminated soon after heating. Although botulism is a rarity in cured meats, semi-perishable canned hams and some other cured products are known to support the growth of C. botulinum. The extent of any growth and toxin formation is usually dependent on the relative presence and competitive growth of the normal lactic microflora (Niven, 1961).

The "microbial inversion" that occurs in the normal salting and curing process provided the origin for all fermented meat products. Early sausage makers in Europe and the Mediterranean countries learned that proper handling of these preserved sausage products yielded products with good taste and stability. Various geographical regions developed unique varieties that varied in size, shape, texture and flavor. Their general acceptance led to the use of specific names for these varieties that were usually derived from the city or area where they were developed. The successful and consistent manufacture of all these products, however, was primarily dependent on the control of the raw materials and the processing conditions. Today, we realize that these environmental controls select for the most desirable lactic acid bacteria that effectively convert the sugars to lactic acid yielding the desired organoleptic characteristics, stability, and safety. Although meat fermentation originated as a "spoilage" condition (i.e. souring), man learned that he could effectively control its degree to provide a stable, nutritious, and widely acceptable food product.

CHAPTER 2
Dry and Semi-Dry Sausages

Definitions

The manufacture of dry and semi-dry sausage traditionally has been more of an art, than a science. The manufacturing principles and techniques were passed from one generation to another with very little documentation, and minimal literature is available on the subject (Figure 1). As of this writing, the United States Department of Agriculture had not formally defined the categories for Dry and Semi-Dry Sausages (USDA Meat and Poultry Inspection Regulations Subparts I and J), as has been done with the four other types of sausage products (USDA, 1973).

One classification generally employed for dry and semi-dry sausage products, manufactured in the United States, is based on ethnic origin, meat formulation, and process features (Table 8). The semi-dry, or Germanic varieties, originated in Northern Europe and traditionally were smoked and often cooked. They contained either beef, and/or pork and beef, and were lightly spiced. The Italian varieties, or dried sausages, originated in Southern Europe, were predominately pork, heavily spiced, and not smoked or cooked. Lebanon bologna is an all-beef product which is heavily smoked but not cooked. It is a unique semi-dry type of product and is common to the areas surrounding Lebanon, Pennsylvania (Table 9).

The meat products also can be characterized, and indirectly are regulated in the United States, by final moisture content and/or moisture to protein ratio (Table 10). Dry sausages have moisture to protein ratios of 2.3:1, or lower. This group includes both the Italian and Germanic varieties such as Italian Salame (1.9), Genoa salami (2.3), pepperoni (1.6), and Hard Salami (1.9). Final moisture content ranges from 25-45%. Semi-dry sausages have moisture to protein ratios in excess of 2.3:1 and they are generally Germanic in origin (Summer Sausage, Thuringer, pork roll, etc). Final moisture content ranges from 40 to 50%. Lebanon bologna is unique in that it has a high moisture content (55-60%), very low pH, and high sugar and salt content.

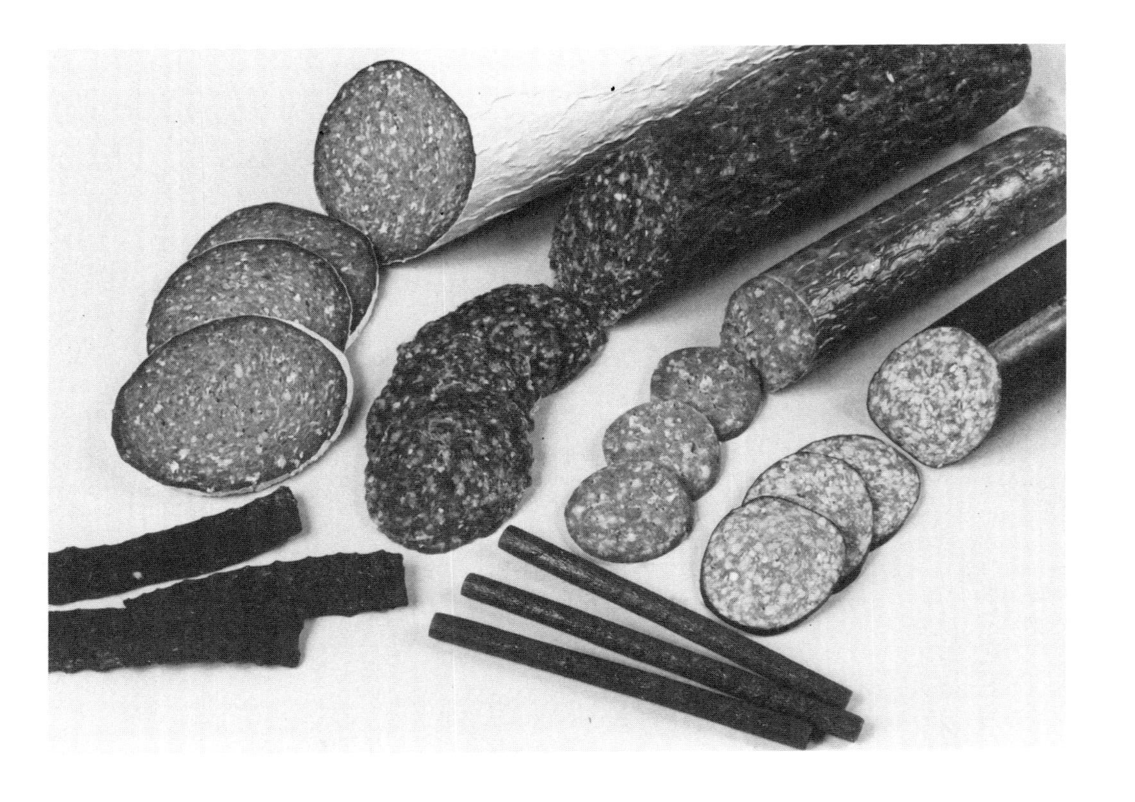

Figure 1. Fermented sausages are merchandised in a variety of sizes, shapes, and colors.

Table 8. Processing features--dry and semi-dry sausages (Terrell et al., 1978).

Type	Process	Meat block	Process features
Germanic	Smoked/cooked	Beef/pork Summer sausage Thuringer Cervelat	18-48 hr. @ 90-100F (5-9 da. w/o culture) Cook to 137F (50% final moisture)
Italian	Dried	Pork/beef Genoa salami Hard salami Pepperoni (San Francisco style)	18-48 hr. green room 30-60 da. dry room Decrease R.H. over time (35-40% final moisture) mold on surface
Lebanon	Smoked	Beef Lebanon bologna	10 da. salt @ 40F Use KNO_3 cure Cold smoke @ 110F 4-8 da. (50 + % final moisture)

Table 9. Moisture to protein range--dry and semi-dry sausages (adapted from Terrell et al., 1978).

Product	Moisture-Protein (Range)
Bologna, Lebanon	2.0-3.7 to 1
Capicola	1.3 to 1
Cervelat, dry	1.9 to 1
Cervelat, soft	2.6 to 1
Pepperoni, dry	1.6 to 1
Salami, dry	1.9 to 1
Salami, soft	2.0-3.7 to 1
Summer sausage/Thuringer	2.0-3.7 to 1
Jerky	0.75 to 1

Table 10. Composition of dry and semi-dry sausages (from Terrell et al., 1978).

Analysis	Thuringer Cervelat, Summer sausage	Genoa salami	Pepperoni	Lebanon bologna	Pork roll
Moisture	50	36	30	56	45
Fat	24	34	39	16	34
Protein	21	22	21	22	17
Salt	3.4	4.8	4.2	4.5	3.6
Sugar	0.8	1.0	2.4	4.1	2.2
pH	4.9	4.9	5.0	4.7	4.8
Total acidity	1.0	0.79	0.4	1.3	1.0
Yield	90	70	64	93	87

Recently, major manufacturers of fermented sausage products in the United States have proposed guidelines for Good Manufacturing Practices for Dry and Semi-Dry Sausages (AMI, 1982b). They have defined dry sausages as "chopped or ground meat products that, as a result of bacterial action, reach a pH of 5.3 or less and are then dried to remove 25 to 50 percent of the moisture to have a moisture to protein ratio no greater than 2.3 to 1.0". Semi-dry, fermented sausages are defined as "chopped or ground meat products that, as a result of bacterial action, reach a pH of 5.3 or less and undergo up to 15 percent removal of moisture during the fermentation and heating process. In general, the semi-dry sausages are not subsequently dried in a dry room but they are packaged soon after the fermentation/heating process is completed. They are generally smoked during the fermentation cycle and have moisture to protein ratios no greater than 3.7 to 1.0".

The proposed Guidelines additionally state that this pH reduction is attributed to the action of lactic acid-forming bacteria which may be added to the meat by either a commercially prepared starter culture or a "back inoculum" from a previously fermented "mother batch". Although fermentation can also be initiated by the lactic acid bacteria occurring naturally in the meat, the Good Manufacturing Practices encourage manufacturers to use more controlled procedures so as to minimize risks.

Dry and semi-dry sausages are manufactured by formulating the meat, spice, and cure components at cold temperatures (20 to 30F, -6.7 to -1.1C), stuffing into the proper sausage casings, incubating the meat at higher temperatures (70 to 110F, 21 to 43C), and subsequently drying the products at temperatures of 50 to 70F, 10 to 21C). Some of the products are also either partially cooked, or fully-cooked, following the incubation stage. During the incubation stage, the product is generally fermented by lactic acid microorganisms reducing the meat pH from 5.8 - 6.2 to 4.8 - 5.3. This fermentation allows the product subsequently to release moisture more uniformly and rapidly. The lactic acid also serves to denature the meat protein resulting in a firmer texture. Inadequate fermentation can result in insufficient drying, "case hardening" of the surface, soft product, a surface "ring", and product "collapsing" (Figures 2 and 3). It also can result in product spoilage, off-flavors, and a potential health hazard since the initial pH reduction serves to preserve the product and yield a desirable flavor.

Meat proteins

The myofibrillar, or structural, proteins are responsible for the binding of water in meat. Water-protein interactions and protein-protein interactions determine the size of the spaces in which the water molecules are harbored in the protein network. The amount of water immobilized depends on the space available between the three-dimensional network of the filaments. In fresh meat, the minimum water-retaining capacity is observed around pH 5.0-5.1 which corresponds approximately to the isoelectric point of the fibrillar proteins in the normal ionic environment of meat (Figure 4). At the isoelectric point, the fibrillar proteins have a maximum of charged groups on their surface and thus have a maximum of hydrophility. This isoelectric condition exists when the number of positive charges is equal to the number of negative charges. When the pH is above the isoelectric point, some of the positive charges are removed, and there exists a surplus of negative charges. This condition results in a repulsion of the filaments--leaving more space for the water molecules. In fresh meat, the same result is observed at a lower pH than the isoelectric point.

The addition of salt to meat affects the total and relative number of charged groups on the filaments. Sodium chloride addition increases the water retaining capacity (i.e. binding) and the swelling of meat on the alkaline side of the isoelectric point (Swift and Ellis, 1956). The chloride ion is bound strongly by the positive charged groups and they are practically eliminated. The sodium ion is bound only weakly by the negative charges, and the net effect is a displacement of the isoelectric point toward a lower pH (Niinivaara and Pohja, 1954) and an increased space between the filaments at or above pH 5 (Figure 5).

When meat tissue (generally containing salt) is fermented, the resulting lower pH allows water molecules subsequently to be released more readily and uniformly since the ultimate pH is closer to the isoelectric point. In addition, the lower pH "denatures" the meat protein whereby the native protein structure or conformation (required for biological function) is altered with the unfolding of the polypeptide chain(s). As the denaturation process continues, the

Figure 2. Case hardening and collapsing are undesirable conditions resulting from inadequate fermentation.

Figure 3. Case hardening can often result in non-uniform drying and internal product voids when the moisture finally "breaks out".

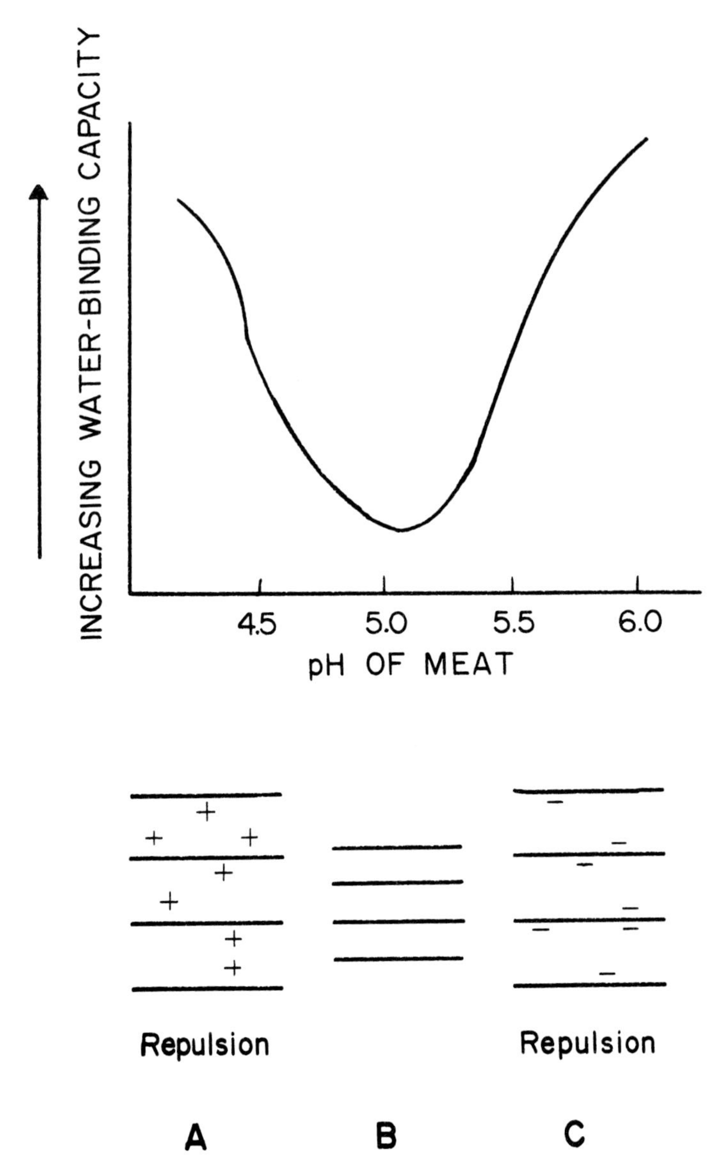

Figure 4. Effect of pH on amount of immobilized water in meat through its impact on the distribution of charged groups on the myofilaments and size of space between them.

(A) Excess positive charges on the filaments.
(B) Balance of positive and negative charges.
(C) Excess negative charges on the filaments.

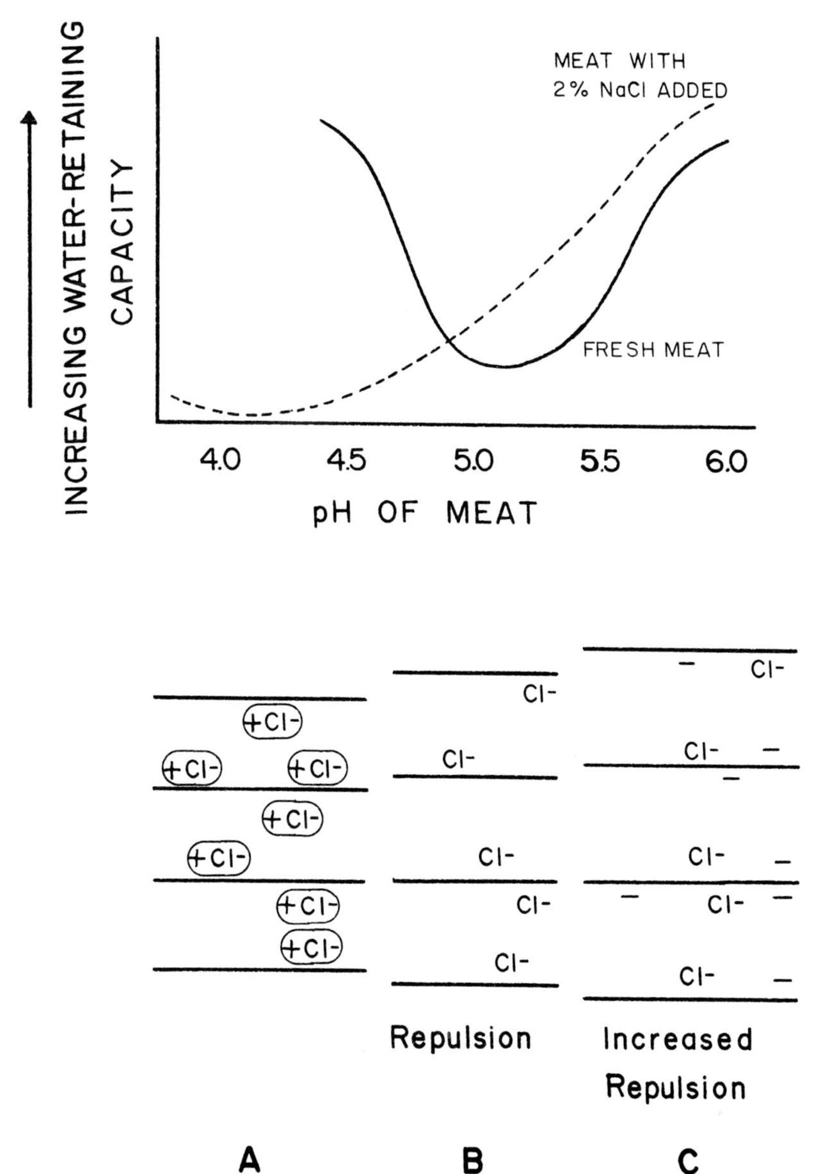

Figure 5. Modified distribution of charged groups on the myofilaments through addition of NaCl and its effect on the amount of immobilized water in meat at various pH values.

(A) Balance of positive and negative charges through Cl- annihilating excess positive charges.

(B) Excess negative charges through binding of Cl-.

(C) Increased excess of negative charges through binding of Cl-.

extensive unfolding exposes side chains which attract one another with the consequent formation of aggregates of increasing size. Ultimately, these aggregates reach a size that they can no longer remain in solution and thus precipitate. This total process is referred to as coagulation, and it is a manifestation of the extensive denaturation initiated by the lower pH (i.e. lactic acid). The denaturation and coagulation of the meat proteins also contributes to the moisture loss, and all three factors contribute to the product "firmness" that is characteristic of most fermented sausages.

Raw materials

Only the best raw materials are preferred in the manufacture of dry and semi-dry sausages. Fresh meats with no microbiological or chemical age, and meats that are well trimmed of sinews, gristle, blood clots and glands are the most desirable. Since these sausages deliberately are "aged" in the incubation and drying stages of the process, any undesirable raw meat characteristics, or defects, can become magnified in the final product. Raw meats with excessive microbial numbers can cause problems in the subsequent incubation phase where undesirable bacteria and yeasts may proliferate. This can result in off-odors and flavors in the sausage and/or textural problems resulting from gas production and/or proteolytic activity. Raw meats, especially frozen materials that have been handled or aged improperly, demonstrate oxidative rancidity that can result in extremely rancid and off-color final products, since the oxidation process is irreversible and accentuated in the drying stage. Any blood clots, glands, and sinews in the raw meat become very apparent in the final product since many of these sausages are coarse-ground and not cooked, so as to dissolve and/or emulsify these defects. Glandular materials also can serve as a source of undesirable, spoilage microorganisms, enzymes, and pathogenic bacteria (i.e. *Staphylococcus aureus*).

Many processors desire lower pH raw materials since the drying process is more effective at reduced meat pH. The desired pH for beef is 5.8 and for pork 6.0. Meats with higher pH values take longer to ferment and tend to hold more moisture.

Formulation

The formulation phase of the process involves grinding, mixing and/or chopping of the meats. A good, clean cut is essential so as to avoid "smearing" the fat that will form a film over the lean tissue, and thereby reduce its ability to lose moisture. In addition, good dry sausages exhibit a clear distinction between fat and lean particles. To achieve the proper cut, the meats must be cold during the formulation phase. If fresh meats are employed, "preconditioning" is recommended to yield 24-28F (-4.4 to -2.2C) meat temperature prior to breaking, grinding, or chopping. Frozen meats are commonly utilized to control formulation temperatures, but their quality should be tightly controlled.

Most processes involve an initial "breaking" of the meat by either a grinder or frozen block cutter. Lean beef materials are generally ground through a 1/8" plate while fat materials and pork are ground through a 3/4" or 1" plate to avoid smear. The various meat materials then are formulated to the desired fat content in a chopper (i.e. silent cutter) or mixer (Figures 6 and 7). If a chopper is employed, the knives must be well maintained to effect the desired cut. In this method, the lean meats are generally chopped first and fat added later. The remaining dry materials as salt, spices, and cure are added last. Paddle mixers (dual shaft) are preferred in a mixing-type operation so as to avoid smearing the fat (Figures 8 and 9). Mixing time should be limited to the minimum that will achieve good distribution, and mixers should not be overloaded. The resulting product mix of meat and dry ingredients from either a chopper or grinder is often reground through a 1/8" to 3/16" grinder plate to achieve the desired particle size.

Processing

In the "traditional process", the sausage mixture is then placed into curing pans, where it is tightly packed in layers of 6 to 8 inches and held 48 to 72 hours at 40 to 50F (4.4 to 10C). During this curing period, the nitrate (i.e. saltpeter) in the mix is converted to nitrite by the nitrate-reducing bacteria, predominately micrococci and staphylococci. The nitrite subsequently enters into the curing reaction with the meat pigments to produce the typical red cure color and the cured meat flavor. The traditional process was developed prior to the knowledge of nitrite and the curing reactions. Various sausage makers learned that "pan curing" was the most effective means to achieve a desirable color and flavor. We now realize that this method allowed the naturally occurring, nitrate-reducing bacteria to produce sufficient nitrite prior to being inhibited by the subsequent, lactic fermentation that would follow at higher temperatures.

The sausage mix subsequently is stuffed into the desired casings (typical of the particular sausage) that hold the product during the remainder of the process and achieve the final product size and shape (Figures 10 and 11). The first sausage casing was probably part of the intestinal tract of the animal that supplied the meat. The choice of containers was fortunate since these natural casings are basically collagen and have special characteristics beyond their unique appearance. During processing, the natural casing in its wet state is permeable to smoke and moisture. Like all collagen, it is hardened and rendered less permeable through drying and the application of smoke (i.e. acidic). Moisture and heat tend to make the casing more soft and porous. As a result, the smoking and cooking processes must be controlled with regard to humidity. If not properly dried prior to smoke application, the smoke will be deposited under the casing and the final color will be pale and dull.

Figure 6. Many fermented meat processors prefer a chopper (i.e. cutmix) to prepare their product. Essentially, this machine is a rotating bowl with cutting blades.

Figure 7. In the chopper, the coarse meat is reduced to the desired size prior to stuffing.

Figure 8. Paddle-type blenders are preferred in a mixing operation for dry and semi-dry sausages.

Figure 9. Paddle-type blenders minimize fat smearing during the mixing operation.

Figures 10 and 11. A meat stuffer inserts the sausage mix into the casings that achieve the desired size and shape of the final product (photo courtesy of Robert Reiser, Inc.).

The tenderness of animal casings varies. Sheep casings, which are used on most small diameter products, are the most tender and are used where the casings are eaten with the product. Hog casings are less tender and are somewhat larger in diameter. They are also eaten with the product. The larger beef casings, in addition to the larger hog bungs and hog stomachs, are removed normally from the product prior to eating. Natural casings are either dried, salted, or packed in a salt slush, and therefore require flushing prior to use.

Currently, dry and semi-dry sausages are processed in either natural casings, regenerated collagen casings (from the corium layer of beef hides), or fibrous casings which are cellulose reinforced with a fiber material for strength. Collagen casings allow the manufacturer to better control product size and machineability. Tenderness can also be controlled and in some products a color, curvature, string ties, or lack of absolute uniformity can be fabricated into the casing. Small diameter collagen casings (i.e. fresh sausage-link types) are designed for the tenderness required since no heat processing will act to increase tenderness. Collagen casings for small diameter smoked product are designed to be tenderized during subsequent heat processing. Drying steps, smoking and humidities in the smokehouse, therefore are critical. During heat processing and smoking, the casing is toughened by drying and smoking. Subsequent cooking with high humidities softens the casing and renders it more tender.

Fibrous cellulose casings are used when size control is critical for subsequent slicing and packaging. Sizing machines can control stuffing to uniform diameters. For sliced product, it is also desirable to minimize the number of ends (i.e. rework). As a result, long lengths are used and the fiber reinforcement is critical for strength. Since most sliced products must be "peeled" prior to slicing, the casing can also be treated with a release agent that allows it to peel easily. In dry and semi-dry sausage, a protein coating on the casing allows the casing to shrink with the product as it dries. To enhance drying in larger diameter casings and to eliminate accumulation of surface air, small holes are "prestuck" into the casing. Fibrous casings require soaking in warm water prior to stuffing to develop their pliability.

The sausage mix should be stuffed at approximately 28 to 30F (-2.2 to 1.1.C) so as to avoid smear from the friction of the stuffing machinery.

In the manufacture of Lebanon bologna, the beef trimmings are salted and held for 10 days at 40F (4.4C). The aged beef is then ground, mixed with sugar, salt, spices, and sodium or potassium nitrate as the only curing agent. The product is then stuffed and smoked for 4 to 8 days during which the fermentation takes place. The heavy smoking imparts a definitive odor and flavor to the product.

Fermentation

After stuffing, the dry and semi-dry sausages are hung on sticks which are generally racked and placed in the "green room", "drip room", or "maturing room", where the fermentation cycle is initiated (Figure 12). The environmental conditions can vary widely, but traditionally, a temperature range from 60 to 75F (15.6 to 23.9C) with 80-90% relative humidity is maintained. The initial humidity is a result of the moisture release from the cold meat as the temperature rises (Figure 13).

The temperature and relative humidities in this fermentation cycle can vary from processor to processor. In the manufacture of most semi-dry sausages, the product is also smoked during this phase of the process. The natural smoke tends to raise the room temperature, and many processors employ "pit houses" where temperature is dependent on the burning sawdust one level below the hanging sausage. Typically, product temperatures are in the range of 90 to 110F (32.2 to 43C). As a result, the fermentation temperatures at the same processor can vary from season to season depending on the weather. In addition, the "pit houses" do not introduce humidity into the room, thus the process humidity is solely dependent on the product moisture that is released, and the climatic conditions. In practice, a pit house full of product will demonstrate high humidities during the beginning of the cycle, and the humidity will be reduced as the cycle progresses. The high moisture and smoke at the beginning of the process also can tend to form a "surface skin" on the sausages, as the acidic smoke deposits on the meat surface and coagulates the protein. This definitely can retard drying, where a subsequent drying room is also employed. In more modern processes, the Summer Sausage is fermented and smoked simultaneously in a controlled atmosphere house where the temperature and humidity are tightly controlled and monitored. These processors generally experience a more rapid fermentation since the product does not tend to dry out during the initial cycle. Surface skin formulation is also reduced. After sufficient fermentation, usually 12 to 24 hours during which the pH falls to 4.8 - 4.9, the product is dried during the subsequent cooking phase. Heating further coagulates meat proteins and pasteurizes the product by inactivating many of the bacteria. Semi-dry sausages generally exhibit product yields of 85-90% of raw weight (i.e. 10-15% shrink).

In the modern manufacture of dry sausages, the product is held anywhere from 70 to 100F (21.1 to 37.8C) with an 80-90% relative humidity. Any natural smoking of the product generally follows the fermentation stage. Liquid smoke, introduced into the formulation, is used frequently to avoid the problems encountered with natural smoking. Control of relative humidity and air circulation are more critical in the manufacture of dry sausages to avoid non-uniform surface drying.

Figure 12. After stuffing, the fermented sausages are hung on sticks which are racked and placed in the fermentation chamber (photo courtesy of Meat Industry Magazine).

Figure 13. Uniform positioning of the sausages in the green room and dry room is essential to provide uniform exposure to the temperature, humidity, and air circulation (photo courtesy of Meat Industry Magazine).

Heating and/or Drying

Following the fermentation stage, dry and semi-dry sausages are either fully-cooked, partially cooked and/or placed directly into a dry room. Generally, semi-dry sausages, such as Summer Sausage, are further heat processed to a range of ultimate product temperatures from 110F to 165F (43C to 74C), depending upon the desired product characteristics. The rate of temperature increase can vary, and the fermentation may continue depending on the growth characteristics of the particular bacteria, existing pH, carbohydrate level, and heat penetration (i.e. casing diameter).

Traditionally, dry sausages go directly from the fermentation room into a drying room without additional heat being applied. Dry rooms can range from 50 to 70F (10 to 21.1C) with 65-75% relative humidity. The control of moisture in the fermented sausages is dependent upon meat particle size, casing diameter, drying air velocity, humidity, pH and protein solubility (USDA, 1977). These various parameters can be combined under two properties, vapor pressure and water binding capacity. The casing diameter, meat particle size, humidity and air velocity will determine the differential vapor pressure between the sausage and the atmosphere. The meat particle size, pH, and the amount of soluble proteins will determine the degree of water binding in the particular sausage. The interaction of these parameters determines the rate and extent of moisture loss in the drying process (Figure 14).

Drying rooms are required that provide uniform temperature and humidity conditions with relatively low air speeds. The problem with attaining these objectives increases with the size of the drying rooms. The lower air speeds make it difficult to control a uniform air flow. In addition, the air at low speed traveling over a long distance may pass through a great mass of product. The desired air conditions will change as moisture is absorbed from some products and carried to others. Different humidity conditions can exist in the same room which can yield "case hardening" in some products (due to excessive drying rates) and mold growth on other products (due to excessive humidity).

The air passing over the product absorbs moisture from the product surface. If this rate of moisture loss is too rapid, "case hardening" occurs where the surface is sealed and the interior moisture is "trapped" within the product. If case hardening occurs in the initial part of the process, the product may not attain the reduced water activity necessary for stability. Quality problems will occur if case hardening occurs in the later stages of the process since a hardening appears only around the surface and a soft inner center may result. If the inner moisture finally "breaks out" with prolonged drying, the moisture release will be non-uniform and product "collapsing" will occur where large surface "grooves" are apparent. To be effective, the drying process must be uniform in that external and internal portions of the sausage lose moisture at the same rate.

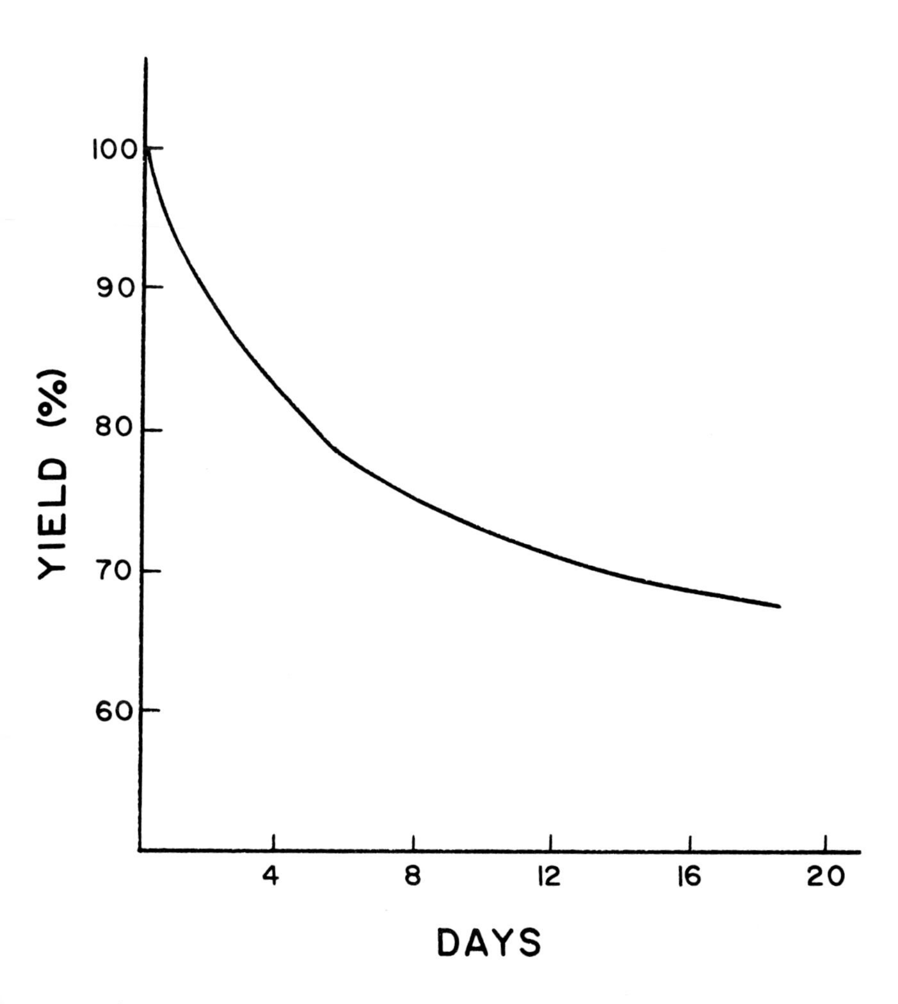

Figure 14. Dry sausage yield-salami

80F, 26.7C (90% RH) - 16 hours
90F, 32.2C (90% RH) - 4 hours
100F, 37.8C (80% RH) - 2 hours
110F, 43.3C (75% RH) - 2 hours
130F, 54.4C (75% RH) - 4 hours
52F, 11.1C (72% RH) - 20 days

The surface "pores" in the meat tissue must remain open allowing internal moisture to escape. Excessive drying rates irreversibly seal these pores.

Mold growth on the product surface is a common observation and often desirable in the manufacture of dry sausages (Figure 15). The control of relative humidity and temperature generally regulates the extent of mold growth. Drying conditions where the relative humidity exceeds 75% usually result in mold growth. Higher drying temperatures also favor rapid mold development. In many types of Italian and Hungarian salami products, specific white and blue-white mold growth is desirable for product flavor and appearance. Drying conditions are usually 65 to 70F (18.3 to 21.1C) with 75-80% relative humidity. Higher humidities can create problems since undesirable molds may occur and sporulation can result where black spores yield undesirable surface color. Proteolytic molds can also "attack" the sausage casing and destroy its integrity. Varying rates of fermentation and ultimate product pH also can result from non-uniform humidity conditions since higher humidities will enhance any microbial fermentation in the dry room. In many dry sausages that are not cooked, the drying process is partially responsible for inhibiting the fermentation. Large drying rooms can be effective with proper duct work that controls the uniformity of air flow. The shorter the distance the air must travel at low speed, the easier it becomes to control the uniformity.

The evaporation rate, therefore, must be controlled to yield desirable product characteristics. Ideally, the rate of moisture loss from the surface should equal the rate of interior moisture migration to the surface. Osmosis will tend to equalize moisture content throughout the product. Control of air speed and relative humidity controls the rate of moisture loss. Sausage with medium size diameter should exhibit 1.0% to 1.5% weight loss per day in the green room. Air speed in the curing or green room is normally held at 1.65 to 2.6 ft/sec. while lower air speeds (0.15 to 0.3 ft/sec.) are suggested in the drying room. The daily weight loss should not exceed 0.7% in the dry room. European processors with lower temperature fermentations generally suggest about a 10% to 12% weight loss prior to transferring product to a dry room. In addition, they have found that a uniform evaporation rate is obtained at a difference of 3-5 points between the water activity (Aw) and the relative humidity (RH) of the room. As an example: Aw 0.93 requires an RH of 88-90%. The RH of the green room and/or dry room must be adjusted downward from day to day as the product dries (Schneider, 1980).

Many European processors gradually reduce humidity throughout the fermentation and drying process whereas most United States manufacturers ferment the sausage under one set of environmental conditions and dry the product under another set of conditions (Table 11). Both methods are effective, and generally the choice of the process and equipment depends on the economic considerations. Processing the sausages in one, controlled-atmosphere room throughout the process

Figure 15. Mold growth on the sausage surface is desirable in the production of Italian Salame (photo courtesy of Gallo Salame).

Table 11. Alternative procedures for making dry sausage.

European - Style		
1st day	72-75F (22.2-23.9C)	94-96% RH
2nd day	68-72F (20.0-22.2C)	90-92% RH
3rd day	65-68F (18.3-20.0C)	85-88% RH
Drying room	53-59F (11.7-15.0C)	75-80% RH
American - Style		
24 to 48 hours to pH 4.9 - 5.0	80F (26.7C)	90% RH
Drying room	50-52F (10.0-11.1C)	68-72% RH

can effect a rapid drying rate, since relative humidity can be gradually reduced as the product dries. However, this method depends on the total utilization of a very expensive room for each batch and it does not provide the versatility to allow for various products that vary in type of casing, casing diameter, and product formulation. The "two-room process" allows the manufacturer to ferment many products at different times in the same fermentation chamber, then place the products in a larger, less-expensive, common dry room. The various products are removed from the dry room when they reach the desired moisture/protein ratio, and/or firmness, and/or meet the USDA requirements for destruction of *Trichinae spiralis* in non-certified pork (USDA, 1973). Dry sausages exhibit product yields of 50-70% of raw weight (i.e. 30-50% shrink).

Microbiology

The fermentation phase of the process in the manufacture of dry and semi-dry sausages traditionally has been the most misunderstood. This phase of the process was termed "greening" or "maturing" with the process chamber called the "green room" or "drip room". Although early sausage makers did not understand fermentation, pH, and microorganisms, they did realize that something happened during this stage of the process that effected optimum drying and resulted in better product color, firmness, and flavor. Although pH meters were non-existent, the sausage maker could tell when this phase was completed by feeling the firmer texture and observing the redder color and tasting the "tang" of the raw mixture. Lack of control at this early phase in the process could result in subsequent product failures due to undesirable odors, and flavors, as well as the product "collapsing" or drying very slowly in the later drying phase.

The traditional process relied on chance and random inoculation of "wild" microorganisms from the environment. These microorganisms were predominately micrococci, coagulase-negative staphylococci, streptococci and lactobacilli. Various strains of these bacteria were unique to the particular sausage ingredients and the individual processing establishments. The addition of salt and the other curing agents to the raw meat generally favored the growth of these types of microorganisms while inhibiting the raw meat microflora. The further holding of the ground meat at cool temperatures under reduced oxygen tension (i.e. in casings) also favored the development of these microorganisms, as did the smoking process.

Although microorganisms do not have to be growing actively to effect enzymatic reactions, they must be present in extremely high numbers for such "resting cell" metabolism to result in significant changes in meat. As a result, most undesirable microbial changes in meat are a result of growth metabolism from contaminants while desirable changes are effected by both growth metabolism and "static cell" metabolism through the addition of large numbers of desirable microorganisms via starter cultures. The extent of any growth metabolism with starters depends on the initial inoculum level.

Most microbes utilize carbohydrates preferentially as a source of energy for growth. These organisms (typically pseudomonads, yeasts, molds, micrococci) can grow aerobically on the meat surface and can oxidize sugars completely to carbon dioxide and water, or incompletely to organic acids. The end-products of complete carbohydrate oxidation have little effect on the odor and flavor of the product, however, as these microbes gain large amounts of energy from these reactions, they often produce copious quantities of cells that result in slime formation. The slime can retard oxygen transfer and/or if for some other reasons the oxidation is incomplete, the metabolism can become anaerobic with the production of a wide variety of fermentation products. In addition, many microorganisms associated with "salted" and "cured" meats (i.e. lactic acid-type bacteria) are incapable of significant carbohydrate oxidation and must rely upon fermentation for growth and metabolism.

The nature of the fermentation end-products depends primarily on the type of microorganism (Table 12). The homofermentative types produce essentially lactic acid from carbohydrate metabolism, yielding a lower pH and a sour, or "tangy", flavor (Table 13). The heterofermentative types can produce a variety of end-products including ethyl alcohol, carbon dioxide, lactic acid, acetic acid, etc. (Table 14). These compounds can have a pronounced effect on the flavor, odor, and texture of the final product. The "uniqueness" of many fermented meats is dependent on a combination of microbial products that result from various levels of the respective microorganisms and the related effects of processing parameters on their growth metabolism. Minor products of carbohydrate metabolism may also affect the flavor of the meat. Acetoin is produced frequently by some lactic acid bacteria, and the mixture of acetoin (i.e. "buttery") and

Table 12. Major products of the anaerobic metabolism of carbohydrates by bacteria that may grow in cured meat products.

Bacteria	Major fermentation products
Homofermentative lactic acid bacteria (streptococci, some lactobacilli, etc.)	Lactic acid
Heterofermentative lactic acid bacteria (leuconostoc and some lactobacilli)	Lactic acid, ethanol, CO_2
Bacillus species (particularly in the presence of nitrate)	Lactic acid, acetic acid, CO_2
Clostridium species	CO_2, H_2, acetic acid, butyric acid, lactic acid, acetone, butanol, etc.

Table 13. Embden-Meyerhof pathway-conversion of glucose to pyruvic acid, to lactic acid (homolactic fermentation).

	glucose
ATP → ADP	↓
	glucose-6-phosphate
	↓
	fructose-6-phosphate
ATP → ADP	↓
	fructose-1-6-diphosphate (furanose ring form)
	↓
	fructose-1-6-diphosphate (open chain structure)
	↓
	glyceraldehyde phosphate
2 Pi 2 NAD → 2 $NADH_2$	↓
	2-1,3-diphosphoglyceric acid
2 ADP → 2 ATP	↓
	2-3-phosphoglyceric acid
	↓
	2-2-phosphoglyceric acid
	↓
	2 phosphoenolpyruvic acid
2 ADP → 2 ATP	↓
	2 pyruvic acid
2 $NADH_2$ → 2 NAD	↓
	2 lactic acid

Table 14. Heterolactic fermentation.

	glucose		
ATP → ADP	glucose-6-phosphate		
NAD → $NADH_2$	6-phosphogluconic acid		
NAD → $NADH_2$	pentose phosphate + CO_2		
+Pi			
NAD → $NADH_2$ +Pi	glyceraldehyde-3-phosphate	acetyl phosphate	-Pi $NADH_2$ → NAD
	1,3 diphosphoglyceric acid	acetaldehyde	$NADH_2$ → NAD
ADP → ATP	3-phosphoglyceric acid	ethanol	
$-H_2O$	phosphoenol pyruvic acid		
ADP → ATP	pyruvic acid		
$NADH_2$ → NAD	lactic acid		

lactic acid (i.e. sour) can yield a different flavor than lactic acid alone. Many lactics also can produce a polysaccharide (dextran or levan) from sucrose that may affect meat flavor.

Many microorganisms can also metabolize meat proteins, peptides, and amino acids through intracellular or extracellular proteolytic enzymes (Table 15). These proteinases and peptidases are complex enzymes and are not well defined with respect to meat substrate specificity and interactions. They are primarily hydrolytic enzymes resulting in a solubilization or liquefaction of protein. Once hydrolyzed, the proteins can yield a variety of amino acids which are further "attached" by microorganisms yielding a range of degradation products including mercaptans, amines, and fatty acids. These compounds can have a dramatic effect on meat odor and flavor, and the desirability of these reactions in flavor development is dependent on the specific products and the quantities produced. Uncontrolled, these reactions usually yield putrefaction.

A common type of enzymatic attack on amino acids is the oxidative deamination to ammonia and the corresponding alpha-keto acid. Clostria can carry out a coupled oxidative and reductive deamination of amino acid pairs to ammonia, a keto acid, and a fatty acid (Stickland reaction). Another form of amino acid degradation is the decarboxylation to carbon dioxide and the corresponding amine (tyrosine to tyramine, ornithine to putrescine, lysine to cadaverine, etc.). Other microbial enzymes can "split" tryptophan to indole and other products; liberate mercaptans and hydrogen sulfide from cysteine and methionine; convert arginine to ornithine, carbon dioxide and ammonia; and degrade histidine by a variety of metabolic pathways (Lechowich, 1978). The overall effect on a meat product is the production of volatile compounds usually associated with some types of spoilage, although the relative importance of these compounds in yielding a distinctive flavor and "aroma" deserves further study. These chemical changes in meat protein can also affect the meat color through direct reaction of the meat pigments with the metabolic products such as hydrogen peroxide, hydrogen sulfide, and nitrite. Other changes can be initiated through changes in the oxidation-reduction potential produced by the bacterial action.

Some microorganisms can also degrade the meat fats by either hydrolysis via a lipase and/or oxidation by oxidases. These reactions can yield "rancidity", although most rancidity problems in meat products are not of microbial origin (Lechowich, 1978). Lipolytic organisms often encountered in meats are pseudomonads and other gram-negatives, bacilli, yeasts, and molds which also are quite active in the oxidative degradation of the fatty acids. Many of the free fatty acids liberated by hydrolytic cleavage of the fats are inhibitory to microorganisms. The peroxides produced through the oxidation of unsaturated fatty acids also are toxic to many microorganisms, and the inhibitory effects observed with some starter cultures has been attributed to these end-products of the starter's metabolism. The mechanism by which microorganisms oxidize the saturated fatty acids

Table 15. Metabolic systems of microorganisms that attack the protein portion of meat products.

System	Typical products
Proteolysis	Soluble peptides
Deamination of amino acids	Ammonia, keto acids, fatty acids
Decarboxylation of amino acids	CO_2, amines such as tyramine, putrescine, cadavarine, etc.
Metabolism of specific amino acids	H_2S and mercaptans from cysteine and methionine; indole and numerous other products from tryptophan

is B-oxidation which removes 2-carbon fragments. In addition, microbial systems that produce peroxide will catalyze the chemical oxidation of the unsaturated fatty acids in meat.

Although fat and protein degradation is associated most often with meat spoilage, the controlled development of unique, meat product organoleptic characteristics is dependent on the control of many of these degradative reactions via processing formulations, parameters and/or specific microbial starter cultures. Specific strains of mold, yeast and micrococci-type organisms are added as starters for their respective proteolytic and lypolytic activities.

Two important types of microbial contaminants were initially required for the manufacture of dry and semi-dry sausages. One type was necessary for the reduction of the added saltpeter (i.e. nitrate) to nitrite for formation of the cured meat color and flavor while the second microbial type was needed to effect a fermentation of the added sugar and to result in the tangy flavor and product stability (Deibel et al., 1961).

The micrococci and coagulase-negative staphylococci have always been associated with cured meat products, particularly fermented sausages, since they can tolerate high salt concentrations and are somewhat more resistant than other microorganisms to nitrite, smoke, and the drying process. They also reduce nitrate to nitrite and many strains also reduce nitrite. Before the manufacturers realized the importance of sodium nitrite, the microbial reduction of nitrate was extremely important to effect the most desirable product. Since these microorganisms are sensitive to acid, the nitrate reduction needed to occur prior to the subsequent fermentation that would inhibit their growth and metabolism. As a result, "pan curing" was employed whereby the growth of the micrococci was favored by the cooler temperatures and more aerobic conditions in the thin meat layers. The micrococci are aerobic in nature but some will grow anaerobically in the presence of nitrate. Their physiology and other attributes will be discussed in more detail in a later chapter. Although many sausage manufacturers still employ this traditional method, some no longer depend upon a microbial reduction of nitrate, but rather directly add nitrite or a mixture of nitrate and nitrite to the sausage mix. One notable exception is the production of Lebanon bologna where many processors still add nitrate and are still dependent on the microbial reduction.

The second microbial type that effects the desired fermentation has been described collectively as the "lactic acid bacteria". In fermented meats, the natural isolates are primarily lactobacilli. Numerous strains of these bacteria ferment the added sugars primarily to lactic acid, reducing the meat pH. However, other end-products of fermentation can be encountered with particular strains, and the resulting flavors can provide a "uniqueness" to a particular product.

Since the success of the process and product was dependent on a "chance inoculation" with the most desirable microorganisms, the results achieved with the traditional process were extremely variable. Product failures were notably common where the sausage was completely unappetizing and inedible, or actually "exploded" in the process. In between the extremes of ideal and inedible products, many of the sausages were mixtures of good and bad contaminations, where the final product was judged only marginally acceptable.

Backinoculum

Proper sanitation and control of the raw materials, the formulation, and the processing conditions were initially the only means to control the development of the most desirable bacteria. In attempts to maintain consistency, some sausage makers observed that a portion of each batch might be held over after the fermentation phase, and prior to any heating or drying, to "start" the next batch. This method which is still practiced today, became known as "backslopping", and generally it afforded a more consistent fermentation from batch to batch. The "backslopping technique" has been refined by many processors whereby selected batches with the most desirable characteristics are frozen for subsequent use as inoculum (usually 0.5 to 5%) in further production. Some manufacturers have a special formulation that is specifically designed for use as a "backslop" (i.e. an inoculum for other batches). This formulation generally consists of fresh, lean beef plus the normal spicing and curing components. The lactic flora is allowed to develop "naturally" at temperatures from 60 to 70F (15.6 to 21.1C). The batch is closely monitored throughout the lactic development and fermentation to achieve the optimum inoculum, or "seed" material. In the proposed guidelines for the Good Manufacturing for Dry and Semi-Dry Sausages (AMI, 1982b), the sausage processors recognize this "back inoculation" procedure as a method of adding lactic acid-forming bacteria to the raw sausage mix. However, they do state that "a product made by back inoculation, where each batch receives a percentage of the formula from a bred and fermented mother batch, the mother batch shall have attained a pH of 5.3 and shall be monitored on a regular basis for lactic acid-producing bacteria and coagulase-positive staphylococci. The same processing guidelines as outlined in these Good Manufacturing Practices should be employed in the manufacture of the back inoculum.

The "backslopping technique", when properly controlled, can be an effective mechanism to provide fermentation and product consistency. The mixed microbial population often affords more of a "wild" flavor and aroma which may, or may not, be desirable. However, if the "backslop" is recycled on a daily, or even weekly, basis the subsequent inoculated batches of sausage may be well underway before a preliminary judgement can be made of the initial batch from which the "backslop" or "seed" was saved. This is true especially of dry sausage where 30 to 60 days are required before the final product can be evaluated. In addition, a meat inoculum prepared in this way contains numerous strains of lactic and possibly other bacteria. The various microbial populations compete in each fermentation, and the dominance of individual strains will vary, yielding different rates of fermentation and different product characteristics. The recycling process also can promote undesirable as well as desirable microorganisms. Spoilage bacteria and some pathogenic bacteria can adapt to the process and gradually increase in number with each recycling. They may establish themselves to the extent that the product is affected adversely before the sausage maker realizes the problem. Once established in the processing facility and product, the resistant strains are difficult to eliminate. A thorough equipment and plant sterilization generally is required which also eliminates the desirable microorganisms. As a result, the whole selection process must be renewed to achieve any fermentation.

Starter Cultures

The initial concept of a pure, microbial starter culture for the manufacture of dry and semi-dry sausages was formally suggested through patents issued in 1921 (Kurk, 1921), 1928 (Drake, 1928), and later in 1940 (Jensen and Paddock, 1940). Further investigations led to the first commercial starter culture offered to the sausage industry in the United States in 1957 (Deibel and Niven, 1957). Initially, this culture consisted of lyophilized cells of *Pediococcus cerevisiae* (later reclassified as *Pediococcus acidilactici*), on a dextrose carrier. This microorganism does not reduce nitrate to nitrite and, to permit the elimination of the traditional pan-curing step, the formulation change from a straight nitrate cure to a straight nitrite, or mixed nitrate-nitrite cure, was made (Niven et al., 1958).

Today, various strains of lactic acid bacteria, micrococci, and molds are available in frozen and freeze-dried forms for use as starter cultures (Figure 16). These cultures consist of both single microbial strains and mixed cultures (Bacus and Brown, 1981).

The formulation and processing parameters for manufacturing fermented sausages with starter cultures may, or may not, significantly differ from the more traditional techniques. In general, the pan-curing step has been eliminated and the fermentation phase has been accelerated. However, some processors primarily depend on a starter culture to provide a consistent inoculum of the most desirable micro-organisms, and they do not want a shortened fermentation cycle. This philosophy results from a desire to achieve a specific shrinkage (i.e. moisture loss) and/or smoke application during the fermentation phase of the process. A minimum amount of time is usually required to accomplish these objectives.

A "starter culture process" relies on the addition of an "active" culture during the formulation phase of the process (Figure 17). Although most processors add the culture after the blending of the dry ingredients into the meat, some may add the culture to the raw meat first, to achieve better distribution. The raw meats are mixed more effectively prior to the addition of the other ingredients. Salt extraction of the protein will "bind" the mixture and other dry ingredients will "dry out" the blend. In addition, large meat pre-blends are often prepared that are then sub-divided for formulation with other ingredients. An effective culture addition can be achieved when fewer total batches and "culture units" are involved. In any event, it is important to avoid direct contact between the viable microbial culture and the curing components (i.e. salt, nitrite) that may reduce the culture viability and activity. Most cultures are available in a concentrate form, and these are diluted with water to achieve better distribution (Figure 18). Freeze-dried forms also require a rehydration step to maximize effectiveness.

Many manufacturers have altered their fermentation process to maximize the benefits of employing a starter culture. Initially, the available starter culture had an optimum growth temperature (95 to 105F, 35.0 to 40C) higher than was traditionally employed in the fermentation phase of the process, especially with dry sausages (60 to 75F, 15.6 to 23.9C). As a result, the fermentation temperature was increased with a rapid acid production. After attaining the desired pH, the sausage was either subsequently cooked or placed directly into the dry room.

The higher fermentation temperatures created quality and safety problems in the manufacture of some dry sausages. "Greasing problems" often occurred as the pork fat liquefied, and the rapid fermentation could yield a strong acid flavor and odor. Growth and toxin production by some strains of Staphylococcus aureus were encouraged by the high temperatures, high salt level, and lack of smoke. In some cases, these pathogenic staphylococci would predominate in the early

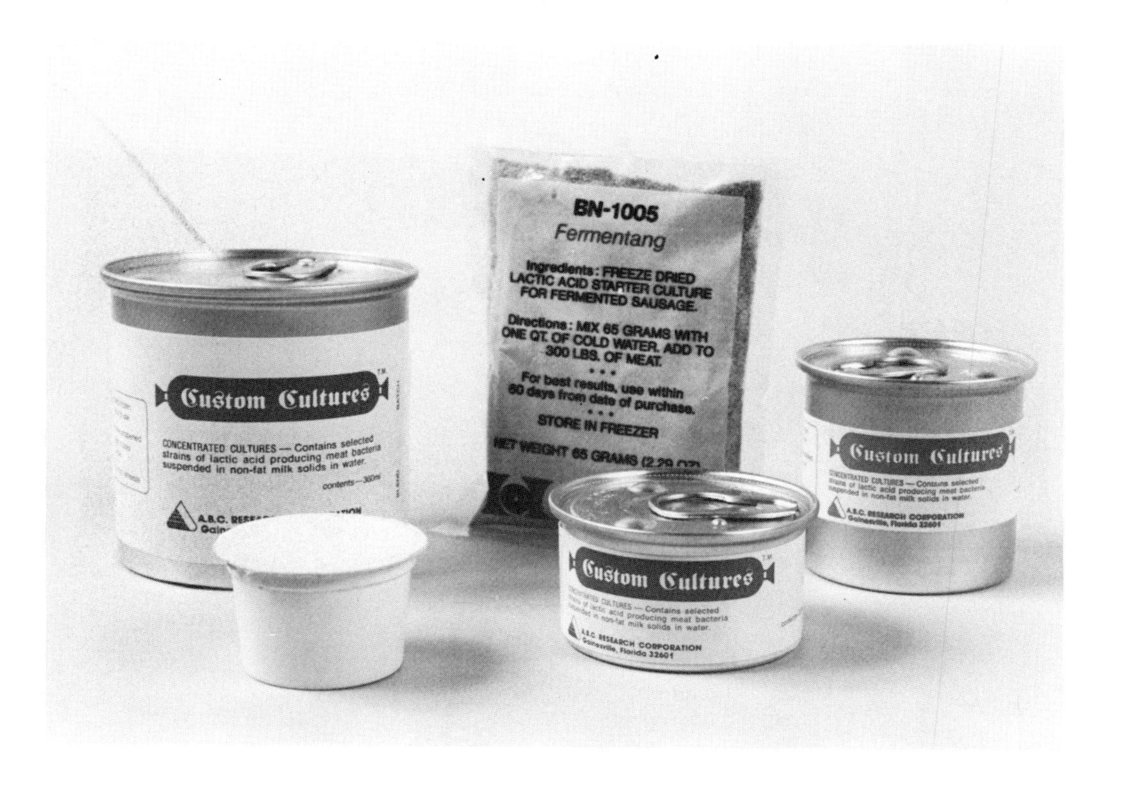

Figure 16. Meat starter cultures are commercially-available in frozen and freeze-dried forms in a variety of containers.

Figure 17. Good distribution of the starter culture is essential to provide uniform performance.

Figure 18. Starter cultures are generally diluted with water to effect better distribution.

stages of the fermentation and would result in unsafe product (USDA, 1977).

As a result, the initial starter cultures primarily were utilized in some semi-dry products where higher temperatures and natural smoking were employed (Niven et al., 1958). The heavy smoke application during fermentation tends to minimize staphylococcal growth at the surface of the product where toxin production is a problem (Barber and Deibel, 1972). Further culture developments resulted in better strains that were effective at lower fermentation temperatures. Currently, a variety of cultures are utilized with fermentation temperatures varying from 50 to 115F (10.0 to 46.0C) depending upon the specific culture, product, and establishment.

Formulations

The following formulations and processes are a few examples of the ingredients utilized and the respective processing parameters for typical dry and semi-dry sausages (Rust, 1977).

<u>Summer Sausage</u> (Semi-dry)

pork and beef, 100 lbs. (should result in a composition of approximately 30 percent fat)

2 lbs. dextrose
3 lbs. salt
2 lbs. cane sugar
¼ oz. sodium nitrate
1/8 oz. sodium nitrite
6 oz. coarse ground black pepper
1 oz. whole mustard seed
½ oz. ground nutmeg
2 oz. ground coriander
½ oz. ground allspice
1-2 oz. garlic powder or equivalent in fresh garlic *Pediococcus* starter culture

The meats should be ground through a ¼-inch to 3/8-inch plate and then mixed thoroughly with the salt, seasoning, dextrose and curing ingredients. However, the product should not be overmixed. After these ingredients are thoroughly blended, the starter culture is added and the ingredients mixed for 3 to 4 minutes or more, depending on the mixer speed. The product is reground through a 1/8-inch or 3/16-inch plate and stuffed in appropriate natural or fibrous casings, usually about 2 inches in diameter.

The following smoke schedule should be used with a starter culture:

	Dry Bulb	Wet Bulb	
16-20 hours	110F (43C)	105F (40C)	1 hour smoke near middle of cycle (to 140F internal temperature)
1½-3 hours	155F (69C)	140F (60C)	
3 minutes	hot shower		

Thuringer Cervelat (Semi-dry)

This is a basic formula for a non-cooked fermented sausage, often called a soft summer sausage.

60 lbs. cow beef (beef hearts may be substituted for up to 15 lbs. of the cow beef)

30 lbs. 80/20 pork trim (certified Trichina free)

10 lbs. 50/50 pork trim (certified Trichina free)

2.8 lbs. salt
2 lbs. dextrose
2 lbs. cane sugar
6 oz. coarse ground black pepper (1 oz. of whole peppercorns may be substituted for 1 oz. of the ground pepper)
1 oz. whole mustard seed
2 oz. ground coriander
½ oz. ground nutmeg
½ oz. ground allspice
¼ oz. sodium nitrate
1/8 oz. sodium nitrite
Pediococcus starter culture

The processing procedures are very similar to summer sausage and the same precautions apply. The meat ingredients should be ground through a ¼-inch or 3/8-inch plate and thoroughly mixed with all the other ingredients except the starter culture. The starter culture then is added and the mixing continued for 3 to 4 minutes.

The product is reground through a 1/8-inch or 3/16-inch plate (1/8-inch preferred) and stuffed in sewed bungs or other appropriate casings.

Where whole peppercorns are used in the formula, the meat should be ground through the 1/8 or 3/16-inch plate before mixing, with the stuffing following directly after mixing.

The following smoke schedule is suggested:

1. 20 hours at 100F (37.8C) - 85% to 90% R.H.
2. 160F (71C) - 30% to 40% R.H. till product reaches 120F (49C) - if Trichina free trimmings are used).

After smoking the product should be showered and then allowed to stand at room temperature for 4 to 6 hours before chilling.

Hard Cervelat (Dry)

Boneless Chucks	26 pounds
Frozen Pork Trimmings	60 pounds
Shoulder Fat	14 pounds
Salt	3 lbs. 12 oz.
Whole pepper	1 oz.
Ground pepper	6 oz.
Red pepper	4 oz.
Sugar	8 oz.
Sodium nitrate	½ oz.
Sodium nitrite	1/8 oz.
Lactobacillus starter culture	

The meat is ground through a 1/8-inch plate and mixed with the curing ingredients and seasoning. After curing for 24 to 48 hours and remixing, the meat is stuffed into beef middles about 11 inches long or hog bungs 25 inches long. The sausage then is hung in the curing cooler for 48 hours, removed from the cooler, warmed up, surface dried and then smoked overnight.

Italian Style Salame (Dry)

Boneless Chucks	20 pounds
Lean Pork Frozen Shoulder Trimmings	48 pounds
Back Fat Frozen Pork Trimmings	20 pounds
Shoulder Fat	12 pounds
Salt	3 lbs. 6 oz.
White pepper	2 oz.
Whole pepper	½ oz.
Sodium nitrate	¼ oz.
Sodium nitrite	1/8 oz.
Fresh garlic (or equivalent in garlic powder)	1 oz.
Lactobacillus starter culture	

To 900 pounds of this sausage is added the following additional seasoning:

Fresh Red Wine	2 quarts
Whole Nutmeg	1
Cloves	1¼ oz.
Cinnamon	½ oz.

The nutmeg and cinnamon are placed in a bag and cooked with the wine for 10 to 15 minutes just below the boiling point. The wine is then strained and cooled. When cool, it--together with the curing material, pepper, and garlic--is mixed thoroughly with the meat after the beef has been ground through a 1/8-inch plate and the pork through a ½-inch plate. The meat is stuffed into casings the size of export hog bungs. The stuffed sausage is hung in the "green room" for 36 hours to dry. After the casing has been dried, the stuffed product is wrapped with twine commencing at the small end, making a hitch of the twine every half inch the whole length of the sausage to the top end. The sausage is then hung in a dry room at a temperature of 50F (10C) for 9 to 10 weeks.

German Style Salami (Dry)

Boneless Chucks	50 pounds
Regular Pork Trimmings	50 pounds
Salt	3 lbs. 6 oz.
White pepper	3 oz.
Sodium nitrate	½ oz.
Sodium nitrite	1/8 oz.
Garlic	1 oz.
Sugar	1 lb. 6 oz.

Micrococcus/*Lactobacillus* starter culture blend

This is a coarse-cut sausage. The beef is ground through a 1/8-inch plate and the pork through a ½-inch plate and mixed with the spices and curing ingredients. After being cured, the meat is stuffed in casings about 3½ inches in diameter and 20 inches long. After stuffing, the sausage is wrapped with twine looped once every two inches and slightly drawn into the casing to give the sausage a scalloped appearance. The sausage is usually dried, but can also be very slightly smoked.

Pepperoni (Dry)

Lean Pork Trimmings	50 pounds
Regular Frozen Pork Trimmings	20 pounds
Boneless Chucks	30 pounds
Salt	3 lbs. 6 oz.
Granulated sugar	4 oz.
Sodium nitrate	¼ oz.
Sodium nitrite	1/8 oz.
Cayenne pepper	8 oz.
Pimiento	8 oz.
Whole aniseed	4 oz.
Peeled garlic (or equivalent in garlic powder)	½ oz.
Coarse ground powder	

Pediococcus starter culture

Pepperoni is named from the pepper content of its spicing. The product is coarsely cut. The pork and beef are ground through a 1/8-inch plate and mixed with the curing ingredients and spices. The meat is often cured in pans 6 inches deep for 48 hours at a temperature of 38F (3.3C). It is then remixed and stuffed into pork casings or the equivalent and twin-linked in pieces 10 to 12 inches long. A little meat is stripped from the broken end of each casing so that there will be about ½-inch of casing left to fold up against the side of the link of sausage. The linked sausage is hung in dry rooms for three to four weeks. If dried at too high a temperature, there is danger of the pork fat running from the heat and becoming rancid.

CHAPTER 3
Meat Cultures

Description

Meat starter cultures are fresh, frozen, or freeze dried single or mixed cultures of selected strains of microorganisms with definite characteristics that are beneficial in the manufacture of certain meat products. Commercial starter cultures are now employed by the majority of manufacturers of dry and semi-dry sausages in the United States. These cultures are added to the raw meat to better ensure product safety, shorten fermentation schedules, and achieve unique product quality, consistency, and shelf-life. The meat starter culture has provided the sausage maker an additional control mechanism to ensure that the product is as desired. Although sausages made with starter cultures generally have smaller risks involved, they are not completely without any risks. Technological and hygienical care must still be taken in the preparation of the raw materials, the fermentation, and the subsequent smoking, cooking, and drying of the sausage. The use of a starter culture is only one factor in the successful manufacture of the product. Product failures and inconsistency can also result from lack of control of the other formulation ingredients and processing parameters.

Starter cultures are microorganisms involved in the normal, natural "ripening" process of the sausage mix. They are grown under closely controlled conditions in a liquid medium, concentrated into a smaller volume, and then placed in frozen storage, or other suitable medium (i.e. lyophilization) to preserve their viability and "activity". Cultures are added routinely to the sausage mix in order to achieve a minimum microbial count of one million organisms per gram of product. This initial microbial level generally precludes any dominance of undesirable microorganisms. Each starter culture manufacturer has proprietary methods to achieve the desired purity, yield, biological activity, and stability of their respective products. Research in this area of culture media, growth parameters, equipment, and preservative methods has been well documented (Porubcan and Sellars, 1979). Although this chapter will discuss the predominant starter culture species that are commercially available for use in meat products, a wide variety of microbial species have been investigated (Table 16).

It has been postulated that the requirements of a good meat starter culture are as follows (Deibel, 1974):

a. salt tolerance

b. fast growing in 6% brine (% salt/% moisture x 100%)

c. ability to grow well in the presence of 80-100 ppm nitrite

d. optimum growth at 90F (32.2C) with range from 80-110F (26.7-43C)

e. homofermentative, producing only lactic acid from dextrose

f. non-proteolytic

g. non-lipolytic

h. not produce off-flavors as by-products of fermentation

i. non-pathogenic

j. inactivation around 135-140F (57.2-60C)

Originally, the predominant microflora of processed fermented sausages consisted of lactobacilli, with some pediococci, streptococci, and micrococci (Table 17). Although the raw sausage mixes contained some pseudomonads and other gram-negative rods, these "fresh meat microorganisms" were not detected in the finished products (Deibel et al., 1961).

The development of starter cultures for any fermented food product evolved from isolation and identification of the microorganisms responsible for the desired effect, and eventually their addition to the food at the appropriate stage of processing. As a result, early researchers in fermented meat technology suggested the use of mainly lactobacilli (Jensen and Paddock, 1940) and some nitrate-reducing micrococci (Kurk, 1921), and even some strains of yeast (Coretti, 1977) as starter cultures for the preparation of these sausages. However, the first commercially available (1957) meat starter culture consisted of lyophilized cells of Pediococcus cerevisiae. This strain was selected from 32 strains of lactobacilli and nine strains of pediococci isolated from fermented sausage (Table 18). Apparently, this strain possessed the most desirable physiological characteristics (Table 19) and also was capable of surviving lyophilization and rehydration. The lyophilization technique was judged the best method at the time to afford distribution of a viable culture as reliably and inexpensively as possible. This procedure necessitated the use of a strain that was best able to survive the treatment.

This initial starter culture mainly was recommended for semi-dry sausages such as Summer Sausage, since it had a somewhat high optimum growth temperature and it was found characteristically in processed Thuringer sausage. In 1965, the lyophilized culture was reevaluated

Table 16. Microorganisms as starter cultures for sausages and cured meats (adapted from Coretti, 1977).

A) Bacteria

Fam.: Lactobacillaceae

Lactobacillus

L. plantarum
L. acidophilus
L. casei
L. fermenti
L. brevis, buchneri
Lactobacillus sp.

Streptococcus

Str. lactis
Str. diacetilactis
Str. acidilactici
(Str. faecalis)

Pediococcus

P. cerevisiae
P. acidilactici
P. pentosaceus

Fam.: Micrococcaceae

Micococcus

M. aurantiacus M 53
M. candidus
M. varians
M. epidermidis
M. conglomeratus
M. aquatilis
Micrococcus sp. P 4
Micrococcus sp.
Micrococcus lactis

Fam.: Enterobacteriaceae

Escherichia sp.
Aerobacter sp.
Alcaligenes sp.

Fam.: Achromobacteriaceae

Achromobacter sp.

Flavobacterium sp.

Fam.: Pseudomoadacae

Pseudomonas sp.

Vibrio

V. costicolus
V. halo(de)nitrificans

Fam.: Corynebacteriaceae

Corynebacterium sp.

B) Yeasts

Debaryomyces

D. kloeckeri
D. hansenii
D. canterellii
D. pfafii
Debaryomyces sp.

C) Molds

Penicillium
P. expansum
P. miczynskii
P. simplicissimum
P. nalgiovensis

Scopulariopsis

Scopulariopsis sp.

Table 17. The total viable bacterial counts of some finished sausages (adapted from Diebel et al., 1961).

No.	Bacteria X 10^6/g	Predominant Flora	pH	NaCl (%)	H_2O	Sugar (%)
Summer sausage	48	*Lactobacillus*	5.3	2.5	32.3	0.37
Summer sausage	190	*Lactobacillus*	5.3	2.6	24.9	0.26
Summer sausage	234	*Lactobacillus*	4.9	2.0	48.9	0.66
Summer sausage	362	*Lactobacillus* and *Pediococcus*	4.8	2.6	38.9	0.61
Thuringer	50	*Pediococcus*	5.0	1.7	58.7	0.12
Thuringer	18	*Pediococcus*	4.9	1.6	54.9	0.12
Salami	28	*Lactobacillus*	5.2	3.1	40.5	0.55
Salami	14	*Lactobacillus*	5.4	4.4	27.9	0.66
Cervelat	25	*Lactobacillus* and *Pediococcus*	5.3	3.0	23.0	0.51
Cervelat	14	*Lactobacillus*	4.8	3.3	30.9	0.92
Genoa	38	*Lactobacillus*	4.9	3.8	32.5	0.08
Göteborg	109	*Lactobacillus*	4.8	2.0	45.0	0.41
Lebanon	320	*Lactobacillus*	4.7	2.1	51.0	2.10
Lebanon	0.1	*Streptococcus*	4.7	1.5	63.5	0.27

Table 18. Physiological characteristics of pediococci isolated from fermented sausages (adapted from Deibel et al., 1961).

	No. of Strains		Other
	+	-	
Catalase produced:			
APT broth..................		9	
Glucose agar, 0.05%.........	2	7	
Litmus milk....................			3 acid; 6 no change
Final pH.......................			3.7-3.8
Acetylmethylcarbinol produced..	3	6	
Hydrolysis:			
Sodium hippurate...........		9	
Esculin.....................	9		
Arginine....................	9		
Gelatin.....................		9	
Starch......................		9	
Growth:			
10C.........................	9		
30C.........................	9		
45C.........................	9		
50C.........................	7	2	
NaCl, 5%; 30C..............	9		
NaCl, 10%; 30C.............	7	2	
Fermentation:*			
Lactose.....................	2	7	
Maltose.....................	2	7	
Sucrose.....................	3	6	
Rhamnose....................	2	7	
a-Methyl-d-glucoside........	2	7	
a-Methyl-d-mannoside........	4	3	
Glycerol, aerobic...........	9		
Glycerol, anaerobic.........		9	

*All strains fermented mannose, fructose, galactose, trehalose, glucose, and salicin. None fermented mannitol, dulcitol, sorbitol, adonitol, inulin, sorbose, melezitose, or dextrin.

Table 19. Physiological characteristics of *Pediococcus cerevisiae* strain FP1 (from Deibel et al., 1961).

Characteristic	Result
Motility	-
Catalase:	
APT broth	-
0.05% Glucose	-
CO_2 from glucose	-
Nitrate reduction	-
Litmus milk	No change
Final pH, glucose broth	3.7
Agar shake	Facultative
Acetylmethylcarbinol production	-
Hydrolysis:	
Sodium hippurate	-
Esculin	+
Arginine	+
Gelatin	-
Starch	-
Growth:	
10C	+
30C	+
45C	+
50C	+
5% NaCl, 30C	+
10% NaCl, 30C	+

Xylose, arabinose, mannose, fructose, galactose, sucrose, trehalose, glucose, salicin, cellobiose, and a-methyl-d-glucoside were fermented. Glycerol was fermented only under aerobic conditions. Lactose, maltose, raffinose, inulin, mannitol, melibiose, rhamnose, sorbose, melezitose, dulcitol, dextrin, sorbitol, and a-methyl-d-mannoside were not fermented.

with the subsequent introduction in 1968 of a frozen concentrate (Everson et al., 1970). The lyophilized form had proven insufficient since the required rehydration period was not very practical and yielded inconsistent results. The culture manufacturer had recommended that the stuffed sausages containing the starter culture be held at 80F (26.7C), 85-90% relative humidity, for 12 to 16 hours prior to fermentation to afford rehydration (Everson et al., 1970). Most processing establishments did not have optimum facilities for the fermentation stage of the process and, as a result, processing times, temperatures and humidities tended to change with production schedules, seasons of the year, and even the time of day. Inadequate rehydration procedures enabled "wild contaminants" to predominate while the Pediococcus strain was inactive due to a prolonged lag phase. The frozen concentrate which was made possible by corresponding advances in the technology of storing, handling, and shipping frozen materials eliminated the variable rehydration step. Subsequent refinements in classification techniques also indicated that the actual starter culture strain, originally designated as P. cerevisiae, was more correctly classified as Pediococcus acidilactici (Buchanan and Gibbons, 1974).

Pediococci

The pediococci strains have been the most widely known starter cultures for meat products since the first commercially-available culture was classified as Pediococcus cerevisiae. Early researchers chose P. cerevisiae strain FPI "due to its rapid growth and resistance to lyophilization". They did acknowledge that this starter culture strain possessed characteristics in common with both P. cerevisiae and P. acidilactici, as described in Bergey's Manual of Determinative Bacteriology (1957) and by Pederson in his description of the genus Pediococci (Pederson, 1949). In addition, the specific strain shared characteristics in common with the two species, P. acidilactici and P. pentosaceus, as described in another classification scheme (Nakagawa and Kitahara, 1959). A U.S. patent covering this early work with a Pediococcus starter culture described the microorganisms as P. cerevisiae, but it also embraced P. damnosus, P. perniciosus, P. sarcinaeformis, P. tetragenus, P. acidilactici, Streptococcus damnosus, S. damnosus var. viscosus and S. damnosus var. pentosaceus, and others, "as all being the same species, although classified as severally distinct by earlier researchers" (Niven et al., 1959). The classification of the genus Pediococcus and related microorganisms has been revised since this early work with the subsequent listing of only five species (P. cerevisiae, P. acidilactici, P. pentosaceus, P. halophilus, P. urinae-equi).

The Genus Pediococcus (Genus III) is within the Family Streptococcaceae (Family II) which is classified in Part 14-Gram-Positive Cocci according to the Eighth Edition of Bergey's Manual of Determinitive Bacteriology (Buchanan and Gibbons, 1974). The pediococci are described as "cocci occurring in pairs or in tetrads as the result of alternate division along two perpendicular planes". These microorganisms are gram-positive, non-motile and do not form endospores. They have a fermentative metabolism which is homolactic, producing DL-lactic acid; the L (+) enantiomorph generally predominates. Acid but no gas is produced from glucose, fructose and mannose while sorbitol and starch are not fermented. Nitrates are not reduced to nitrites and gelatin is not liquefied. Their nutritional requirements are complex and they are microaerophilic. Most strains are catalase negative although some possess non-heme catalase activity.

Pediococcus cerevisiae, Pediococcus acidilactici, and Pediococcus pentosaceus are differentiated from the other two species of pediococci by their ability to grow at pH 5.0 but not pH 9.0 (Table 20). Pediococcus cerevisiae is distinguished from both P. acidilactici and P. pentosaceus by its inability to grow at pH 7.0 or at 35C (95F). It also prefers anerobic conditions. The latter two species are differentiated from each other by the ability to grow at 50C (122F). P. acidilactici can grow at 50C (122F) while P. pentosaceus cannot. Both are microaerophilic. Natural isolates of these species commonly are found in spoiled beer and brewers yeasts (P. cerevisiae), sauerkraut and fermenting mashes (P. acidilactici), and a wide variety of other fermenting materials such as pickles, silages, and cereal mashes (P. pentosaceus).

Today, strains of P. acidilactici and P. pentosaceus are commercially available as meat starter cultures (Bacus, 1983; Raccach, 1981). Some strains of P. cerevisiae may occasionally be found in sausage products and may be available as starter cultures, although most of the commercial strains that were originally designated as P. cerevisiae have been reclassified as P. acidilactici. These two classifications have been repeatedly interchanged in the terminology used in promotional literature and in general references to meat starter cultures. The United States Department of Agriculture also recognizes P. cerevisiae in the regulations permitting starter cultures in meat processing (Table 21).

Pediococcus cerevisiae has an optimum growth temperature of 25C (77F) with no growth at 35C (95F). The thermal death point is 60C (140F) for 10 minutes. Characteristically, the organism produces lactic acid but no gas from dextrose and maltose and sometimes galactose and salicin. No acids are produced from pentoses, lactose, sucrose, or mannitol. Diacetyl is generally produced, and some researchers have observed a requirement for carbon dioxide. Niacin and biotin are essential for growth while pyridoxine is stimulatory.

A strain of P. acidilactici was probably the actual starter culture initially available to manufacturers of fermented sausages. Characteristically, this microorganism has an optimum growth temperature of 40C (104F) with a maximum of 52C (122F). The thermal death point is 70C (158F) for 10 minutes, although some strains can be more heat tolerant when recently isolated. Lactic acid is produced from the fermentation of glucose, galactose, arabinose, xylose, salicin,

Table 20. Genus <u>Pediococcus</u> (from Buchanan and Gibbons, 1974).

I. Growth at pH 5.0, not pH 9.0

A. No growth at pH 7.0 or at 35C; prefers anaerobic conditions

1. <u>P. cerevisiae</u>

B. Growth at pH 7.0 and 35C; microaerophilic

1. Growth at 50C

2. <u>P. acidilactici</u>

2. No growth at 50C

3. <u>P. pentosaceus</u>

II. No growth at pH 5.0, growth at pH 9.0; microaerophilic

A. Halophilic

4. <u>P. halophilus</u>

B. Not halophilic

5. <u>P. urinae-equi</u>

and trehalose while some strains produce slight acid from sucrose and lactose. Characteristically, acid is produced from maltose, mannitol, α-methyl glucoside or dextrin. Diacetyl is produced. The organism is extremely fastidious requiring nearly all amino acids except methionine, niacin and biotin. The G + C content of DNA is 44 moles % (Buchanan and Gibbons, 1974). Strains used in sausage processing must also be salt tolerant which is a characteristic that can be lost through repeated transfers in media without salt. An initial description of the strain utilized recommended a minimum salt tolerance of 5% with an essential range of 5 to 8%. The early strains were developed selectively by using laboratory broth cultures containing gradient amounts of salt (Deibel and Niven, 1957).

The relatively high growth temperatures have traditionally favored the use of <u>P. acidilactici</u> in many semi-dry sausages where corresponding fermentation temperatures are employed. This microorganism rapidly ferments dextrose to lactic acid and yields a sharp, "tangy product". The starter culture is used strictly for its "acid-producing capability" and in products where a definitive "acid taste" is desirable, as is common to this class of sausages. Early work with this culture demonstrated that the predominant organoleptic characteristics in summer sausage-type products, aside from the "tangy" flavor, arose from the individual spicing, salt, sugar, and meat components, as well as contributions from the varied processing schedules as practiced by individual plants. Flavor contributions due to secondary bacterial actions (aside from straight lactic acid production) appear to be insignificant in most summer sausage-type products (Deibel et al., 1961). However, these considerations are

Table 21. <u>Meat and Poultry Inspection Regulations</u> (1973) (318.7 Approval of substances for use in the preparation of product).

Class of substance	Substance	Purpose	Products	Amount	
Flavoring agents; protectors and developers.	Harmless bacteria starters of the acidophilus type, lactic acid starter or culture of	To develop flavor.	Dry sausage, pork roll, thuringer, lebanon bologna, cervelat, and salami	0.5 percent	
* *	*Pediococcus cerevisiae*	To dissipate nitrite.	Bacon	Sufficient for purpose.	* *

not totally applicable in other types of fermented sausages (i.e. some dry sausages, Lebanon bologna) as will be described later.

P. acidilactici is still commonly employed as the primary starter culture for semi-dry products, and it is used in some dry sausages that are fermented at temperatures in excess of 90F (32.2C).

In recent years, P. pentosaceus has been promoted as a starter culture for many sausage products since it has a lower optimum growth temperature of 35C (95F) with a maximum of 42-45C (108-113F). The thermal death point has been described as 65C (149F) for 8 minutes. This microorganism produces lactic acid from dextrose, galactose, maltose and usually from arabinose, xylose, lactose, salicin, and α-methyl glucoside. Sometimes slight acid is produced from sucrose while no acid is produced from rhamnose, trehalose, mannitol, dextrin, or inulin. Diacetyl is produced. All amino acids are required for growth, and serine, methionine and lysine are stimulatory. Folic acid, niacin and pantothenic acid are essential while biotin is stimulatory. The G + C content of the DNA is 38 moles %.

Since the earlier pediococci strains available to meat processors, predominately P. acidilactici, were most effective between 80F to 120F (26.7 to 48.9C) the use of lower fermentation temperatures (60F to 80F, 15.6 to 26.7C) resulted in excessive fermentation times. In addition, antioxidants (i.e. BHA, BHT) that are commonly employed in dry sausages were also found to be toxic to P. acidilactici, further inhibiting the fermentation. A selected strain of P. pentosaceus in combination with a stimulatory metal salt, preferably a manganese salt, was recently offered to the meat industry to afford more rapid fermentations at temperatures between 60F to 80F (15.6 to 26.7C). The stimulatory, food grade metal salt added to the culture reduces inhibition that may occur in the final sausage mix, allowing a more rapid fermentation to ensure. Salts that proved stimulatory to the culture include manganese chloride, manganese sulfate, manganese glycoerophosphate, manganese oxide and manganese gluconate and the various non-toxic metal salts of acids which are slightly soluble in water. Other metal ions include ferrous, ferric, magnesium, calcium and zinc although none of these were as effective as manganese. The stimulation of meat starter cultures with metal salts has been previously documented (Chaiet, 1960). Although the P. pentosaceus blend was primarily designed for lower fermentation temperatures, the fermentation range may extend to 120F (48.9C) according to the manufacturer (Raccach, 1981).

Lactobacilli

Before the introduction and subsequent use of the pediococci strains as meat starter cultures, most natural isolates from fermented meats consisted of various species and strains of lactobacilli (Deibel et al., 1961). Lactobacilli are still the predominant microflora in products that are "naturally fermented" from chance inoculation. An early U.S. Patent pioneering the use of a bacterial starter culture in sausage products described the effectiveness of Lactobacillus plantarum, L. brevis, and L. fermente (Jensen and Paddock, 1940). These investigators used several Lactobacillus species to effect the meat fermentation, but they depended upon chance inoculation for nitrate reduction (i.e. curing reaction). Subsequent work with pure cultures failed to achieve successful lyophilized preparations of Lactobacillus isolates with "state of the art" technology, so continuing investigations focused on the Pediococcus strains that were resistant to the lyophilization process.

With the introduction of frozen culture concentrate for sausage in 1968, interest in the Lactobacillus strains was renewed. In addition, the lactobacilli generally have lower growth temperatures which made them more suitable for dry sausage production. In 1974, a patent was issued that described the use of Lactobacillus plantarum, either alone or in combination with other lactic acid producing microorganisms (i.e. P. acidilactici) for preparing dry and semi-dry sausages (Everson et al., 1974). The concept of a "mixed" starter culture was also introduced to afford the sausage manufacturer a broad range of fermentation temperatures.

The specific strain of L. plantarum (NRRL-B-5461) was described as gram positive, non-motile rods, ordinarily 0.6 - 0.8μm by 1.2 - 1 to 6μm, occurring singly or in short chains. The organism was catalase negative and does not grow at either 7C or 45C (44.6F or 113F). The thermal death point is 30 minutes at 63C (145.4F). Glucose fermentation resulted in DL lactic acid with no gas, and the strain was particularly salt tolerant, developing in salt concentrations greater than 9%. A variety of carbohydrates are fermented including fructose, glucose, galactose, sucrose, maltose, lactose, dextrin, sorbitol, mannitol and glycerol. Xylose, dulcitol, mannose, and salicin were among those carbohydrates not fermented. The moles percent G + C content was reported as 43.6 compared to the 45 ± 1 moles % as described for L. plantarum in Bergey's Manual (Buchanan and Gibbons, 1974).

Lactobacillus plantarum is one of 27 described species in the genus Lactobacillus (Genus I) which is within the Family (I) Lactobacillaceae--gram positive, asporogenous, rod shaped bacteria. L. plantarum is generally differentiated from many of the other species by the lack of gas production from glucose, but gas is produced from gluconate. Ribose is fermented yielding lactic and acetic acids, thiamine is not required, aldolase activity and glucose 6-phosphate dehydrogenase and inducible 6-phosphogluconate dehydrogenase are present. The lactic fermentation produces DL-lactic acid.

The organism is a facultative anaerobic and typically, does not reduce nitrate to nitrite. Ammonia is not produced from arginine and milk is acidified. Optimal growth is usually 30-35C (86-95F). Calcium pantothenate and niacin are required. Natural isolates are obtained from dairy products and environments, fermenting plants, silage, sauerkraut, pickled vegetables, and the human intestinal tract. Closely related species are L. casei subsp. pseudoplantarum and L. curvatus (Buchanan and Gibbons, 1974).

Natural isolates described as lactobacilli have been repeatedly recovered as the predominant microflora from fermented meat and sausage products. Most authors have emphasized the occurrence of L. plantarum, but descriptions of L. casei and L. leichmanii have also been documented. In general, the majority of the natural isolates are lactobacilli species variants that do not precisely conform to any specific species, but they are most closely related to L. plantarum. These strains are highly adapted to the specific environment from which they were isolated, and they will out-perform classical strains when both are reintroduced into the sausage environment. L. plantarum NRRL-B-5461 when compared to a known strain (ATCC-14917) in the fermentation of a typical sausage mix developed a lower pH in a shorter time at two different fermentation temperatures (Table 22).

Meat starter cultures composed of lactobacilli are most often utilized in the production of dry sausages where fermentation temperatures are between 60 to 95F (15.6 to 35C). In these types of products, the primary lactic acid fermentation is necessary, but often secondary fermentations and other chemical reactions play a significant role in the organoleptic distinctiveness of the products. Fermentation times range from one to five days in the more traditional dry sausage processes. These lower temperature fermentations for extended times may allow some growth and metabolism by the naturally-occurring microflora resulting in various metabolic end-products that effect the final flavor. Enzymatic reactions in the meat also have more time to be effected prior to cooking, drying, and/or pH reduction. Most dry-products do not have the sharp "tangy" taste that is characteristic of semi-dry sausages. This is especially true with European products where a "mild" taste is preferred, and higher final product pH's (i.e. 5.3-5.6) are more frequently observed (Coretti, 1977).

The wide variety of lactobacilli strains probably contributes to the "uniqueness" of many dry sausage products. Comparisons of commercial Lactobacillus plantarum starter cultures, classical ATCC strains, and various natural isolates of lactobacilli from sausages demonstrate variations in the respective fermentation characteristics, growth temperatures and the rate and type of acid production (Table 23). No appreciable proteolytic or lipolytic activity was detected in any of the strains. All the lactobacilli strains demonstrated the ability to decompose hydrogen peroxide when grown in the presence of myoglobin. This "catalase activity" would definitely retard any oxidation and the destruction of meat pigments that can occur when peroxides are present. Catalase activity has been promoted as a major benefit when employing micrococci as meat starter cultures, and it would appear that lactobacilli would also have this activity when propagated in meat. It has been shown that meat starter cultures of

Table 22. Fermentation rate, Lactobacillus strains (adapted from Everson et al., 1974).

	pH of the sausages aged at 65F (18.3C)		
	Control	NRRL-B-5461	ATCC-14917
Days:			
1	5.80	5.75	5.75
2	5.80	5.68	5.75
3	5.55	5.20	5.45
4	5.22	4.60	4.85

	pH of the sausages aged at 100F (37.8C)		
			L. plantarum
	Control	NRRL-B-5461	ATCC-14917
Hours:			
0	5.80	5.80	5.80
7	5.83	5.80	5.85
17	5.88	5.13	5.75
21		4.60	5.20
24	5.85	4.45	5.10
41		4.45	4.53

lactic acid bacteria are unable to produce and accumulate hydrogen peroxide (Raccach and Baker, 1979), although this characteristic is highly variable, and the observed results are very dependent on analytical techniques.

Most commercial starter cultures are homofermentative, yielding primarily lactic acid from the fermentation of dextrose and sucrose. However, heterofermentative lactobacilli can often occur as natural contaminants and contribute to the sausage flavor through the production of volatile acids, alcohol and carbon dioxide. The role of two heterofermentative lactics, L. brevis and L. buchneri, has been documented and the incorporation of heterofermentative lactics in sausage-making has been suggested (Urbaniak and Pezacki, 1975).

Several strains of heterofermentative lactic acid bacteria (i.e. Leuconostoc mesenteroides) have been isolated from unique varieties of fermented sausages and reintroduced as starter cultures in combination with homofermentative lactobacilli. In one specific

Table 23. Biochemical characteristics of lactic acid bacteria used in dry sausage production[a] (adapted from Nordal and Slinde, 1980).

Strain	API 50L system																			Growth Temp.				
	Glycerol	L-(+)-Arabinose	Mannitol	Sorbitol	Methyl-D-glucoside	Amygdalin	Arbutin-iron-citrate	Esculin-iron-citrate	Salicin	D-(+)-Cellobiose	Maltose	Lactose	D-(+)-Melezitose	D-(+)-Raffinose	Arginine	Glucose, gas Production	Teepol, 0.4%	Teepol, 0.6%	o-Nitrophenyl-β-D-galactosidase	42°C	10°C	8°C	Relative catalase activity[c]	Final pH[d]
ATCC 8014	±[b]	+	+	+	+	+	+	+	+	+	+	+	+	+	-	±	+	+	+	+	+	-	100	4.5
SL	+	+	+	+	+	+	+	+	+	+	+	+	-	+	-	+	+	-	+	+	-	-	121	4.4
DL	-	-	+	+	-	+	+	+	+	+	+	+	+	+	-	+	+	+	+	+	+	-	35	4.8
I-L	-	-	-	-	-	+	±	+	-	-	+	-	-	-	-	±	-	-	±	+	+	+	76	4.9
II-L	-	-	-	-	-	-	-	+	-	+	-	-	-	-	±	±	-	-	-	-	+	+	96	4.6
III-L	-	-	-	-	-	-	-	-	-	-	-	-	-	-	-	±	-	-	-	-	+	+	119	5.3
IV-L	-	+	-	-	-	-	±	+	+	±	+	-	-	-	+	±	-	-	+	-	+	+	92	4.8
V-L	+	+	-	-	-	-	-	±	-	±	±	±	-	-	±	±	-	-	+	-	+	+	156	5.0

[a]The following reactions were positive for all the strains: ribose, galactose, D-(+)-glucose, D-(-)-levulose, fructose, D-(+)-mannose, N-acetylglucosaminem D-(+)-melibiose, sucrose, D-(+)-trehalose, and growth at 13 and 40°C. The following reactions were negative for all the strains: erythritol, D-(-)-arabinose, D-(+)-xylose, L-(-)-xylose, adonitol, methylxyloside, L-(-)-sorbate, rhamnose, dulcitol, meso-inositol, methyl-D-mannoside, inulin, dextrin, amylose, starch, glycogen, acidification of gluconate, urease, catalase, nitrate reduction, acetoin production, and growth at 45°C.

[b]±, Variable.

[c]The activity of strain ATCC 8014 was used as reference.

[d]M_3 medium containing 1.8% glucose, initial pH 6.9, 3 day incubation @ 25 C.

instance, three distinct strains of lactic bacteria contributed to the unique flavor and texture of the product. Two psychrotrophic lactics, one heterofermentative, produced flavor compounds (i.e. ethanol, diacetyl) during the initial precuring stage of the process (3-5 days @ 50F, 10C). Sausage pH decrease was only minimal to 5.3. The product was subsequently smoked at higher temperatures (90-110F, 32.2-43.3C) where the psychrotrophic strains were destroyed, and a homofermentative, mesophilic pediococci strain further reduced the pH to 4.7 to 4.8. The three bacteria were isolated, propagated in the laboratory, and reintroduced as starter cultures in defined proportions so as to achieve the desired floral succession and flavor and acid development. Control of the inoculum proportions and process parameters is extremely important so as to avoid the predominance of one strain. Excessive "flavor development" from the heterolactics in the precuring stage would also result in a swollen, "spongy" product from excessive gas formation. Premature "acid development" from the *Pediococcus* strain would prevent the proliferation and metabolic activity of the psychrotrophs. The use of heterofermentative lactobacilli as starter cultures offers the potential of unique flavor development although process controls must be extremely rigid. This is probably why such strains are not readily available as meat starter cultures.

Micrococci

The role of the micrococci-type starter cultures for meat products is somewhat complex. Unlike the lactic acid bacteria, the micrococci are not added primary to rapidly produce acid and thus, lower meat pH. The micrococci can take an active part in meat curing. Early researchers depended upon chance inoculation by more fermentative bacteria and utilized *Micrococcus aurantiacus* as a starter culture to reduce nitrate to nitrite to effect the curing reaction (Niinivaara, 1955). At about the same time that *Pediococcus cerevisiae* was introduced as a starter culture in the United States, *Micrococcus* strain M53, "or in general suitable strains of the genus *Micrococcus* (Bergey's Manual, 6th ed.)," was made available for European types of fermented sausage. Natural isolates of micrococci were selected based on their ability to reduce nitrate, to form acid from glucose in aerobic conditions, to grow in media containing sodium chloride, and for the intensity of the nitrate reduction (Niinivaara et al., 1964). With the introduction of the straight nitrite cure, or a mixed cure of nitrate and nitrite, the role of the micrococci was diminished, and micrococci are not commonly employed as starter cultures in the United States. However, many European processors still utilize micrococci and combinations of micrococci and lactobacilli as starter cultures to yield their desired product characteristics (Coretti, 1977).

The Family *Micrococcaceae* (Family I, Part 14 Gram Positive Cocci, Bergey's Manual of Determinative Bacteriology, 1974) includes three genera, *Micrococcus*, *Staphylococcus*, and *Planococcus*. These gram positive, catalase positive, salt and nitrite tolerant bacteria appear in clusters of packets of coccoidal cells. The nutritional requirements are variable and all strains grow in the presence of 5% sodium chloride. Many grow in 10-15% sodium chloride. Although twenty-seven different species have been isolated from cured meat products (Niinivaara et al., 1964), the latest scientific classification only provides three separate species for each genus; there seems to be quite a variety of strains with somewhat different characteristics.

The genera Micrococcus and Staphylococcus are of concern in relation to meat products, since Planococcus is generally found in sea water. It is also motile which distinguishes it from the other two, non-motile genera.

Micrococci and staphylococci are differentiated on their ability to anaerobically ferment glucose. Micrococcus species are by definition aerobic, with a strictly respiratory metabolism. Glucose is oxidized to mainly acetate or completely oxidized to carbon dioxide and water. Glucose is metabolized by hexose monophosphate pathway and citric acid cycle enzymes. A functional glycolysis cycle may also be present. The spherical cells occur in pairs or singly. Catalase is produced, and the G + C content of the DNA is 70.7 - 75.5 moles %.

Staphylococcus species are facultative anaerobes; their metabolism being either respiratory or fermentative. Under anaerobic conditions, the main product of glucose fermentation is lactic acid. In the presence of air the main product is acetic acid with small amounts of carbon dioxide. Growth is more rapid and abundant under aerobic conditions. Nitrates are reduced to nitrite, and nitrite is usually reduced by the coagulase-negative strains. The spherical cells occur singly, in pairs and characteristically, in irregular clusters (Gr. n. staphylo. bunch of grapes). Catalase is also produced, and the G + C content of DNA ranges from 30-40 moles %. Staphylococci can be a host for a wide range of bacteriophages which may have a narrow or wide host range. Transfer of characters by transduction has been shown for S. aureus. Most strains grow in the presence of 15% sodium chloride. The staphylococci are also differentiated from micrococci by their sensitivity to lysostaphin (1 unit/ml) and their resistance to lysozyme (100 μg/ml). Micrococci are resistant to lysis by lysostaphin and demonstrate variable sensitivity to lysozyme. Many staphylococci are associated with skin, skin glands, and mucous membranes of warm blooded animals. Pathogenic strains (i.e. S. aureus) are common causing a wide range of infections and intoxications.

The genus Micrococcus consists of three species: M. luteus, M. roseus, and M. varians. These non-pathogenic bacteria are commonly in soil, dust, water and the skin of man and other animals. The nutritional requirements are variable with many strains capable of growth in minimal media. M. luteus and M. varians can produce a yellow pigment while M. roseus produces a red pigment. M. luteus usually produces no detectable acid from carbohydrates and oxidizes a variety of carbon containing compounds completely to carbon dioxide and water. M. roseus yields variable acid production while M. varians definitely produces acid, but no gas, from fructose and glucose. Acid production is variable from galactose, lactose, maltose and sucrose. Nitrate is usually reduced to nitrite by both M. roseus and M. varians but not by M. luteus. The optimum growth temperatures are 25C (77F), 30C (86F), and 22-37C (72-99F), for M. roseus, M. luteus, and M. varians, respectively.

M. varians is commercially available in the United States and abroad as a meat starter culture (Gryczka and Shah, 1979). This strain is characterized as a "poor producer of lactic acid and alone cannot produce acceptable sausage." As a result, it is usually available in combination with a "lactic acid producing meat fermenting bacteria" (i.e. P. cerevisiae). The Micrococcus species (NRRL-B-8048) is beneficial in developing "cure color" in meats in the presence of nitrate and/or nitrite. A problem with many nitrate reducing bacteria is that the nitrate reductase enzymes are pH sensitive (usually 5.6), and rapid fermentation or chemical acidulation will retard nitrate reducing activity. The strain identified reduces nitrate rapidly to achieve sufficient nitrate reduction within the time period that the pH is higher than 5.6 in the sausage. It was also documented that M. varians ATCC 15,306 is not inhibited in its nitrate reducing function until a pH of about 5.2 is achieved, making it also desirable for fermented meats. Bacterial compositions including Micrococcus varians also generate catalase enzyme activity which may retard hydrogen peroxide accumulation. Peroxides in fermented sausages are known to "bleach" the cure color and promote oxidative rancidity (Raccach and Baker, 1978).

Micrococci are commonly employed as starter cultures in the manufacturing of European-style dry sausages (Coretti, 1973). In these slow ripening processes, the micrococci dominate the microflora during the initial phase (up to 4 days), thereafter gradually declining and yielding to the lactobacilli (Figure 19). Most micrococci actively reduce nitrate and nitrite via reductase enzymes. The development of micrococci as starter cultures evolved from a need to control nitrite concentration when a sausage manufacturer employed a straight nitrate cure. Many European processors of dry sausage and Lebanon bologna processors in the United States still use a nitrate cure, depending upon microbial nitrate reduction to "cure" the meat. However, even when nitrite is added, many processors feel that "cure color development" is more effective and more stable when the micrococci are present. This "cure color stability" prevents gray discoloration at the sausage surface and at the cutting face when the product is sliced and exposed to air. This may result from the nitrite reduction to nitric oxide that effects a more rapid curing reaction (Figure 20). In addition, it is often felt that the red color obtained with nitrate, via nitrate reduction, is more intense and stable than the color obtained from the use of nitrite (Anonymous, 1978).

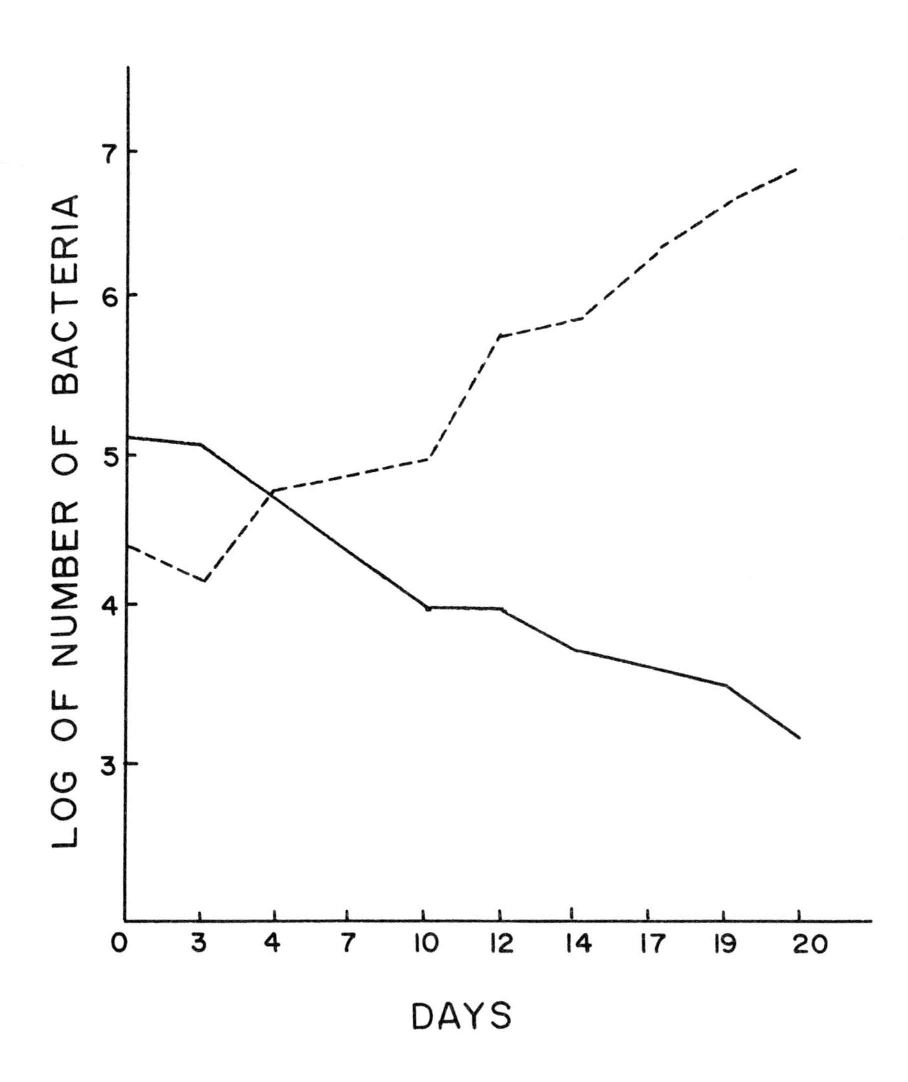

Figure 19. Number of micrococci and lactobacilli during the fermentation of Fruskogorska sausage. ——— micrococci, ------- lactobacilli (Stolic, 1975).

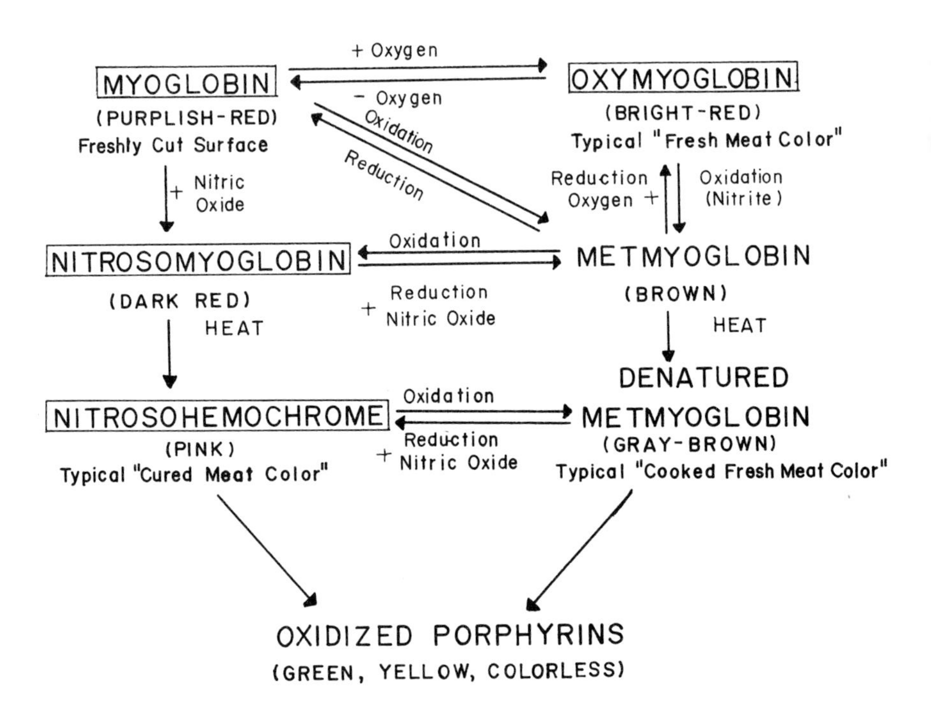

Figure 20. Color changes in cured meat.

Practical experience has also demonstrated that a darker, "richer" red color is obtained in dry sausages, whether nitrate, nitrite, or both are used, when the products are not exposed to further heat processing after fermentation. This observation probably results from the formation during heating of nitrosohemachrome, the light pink pigment typical of cured meat color. Its precursor, nitrosomyogloblin--a dark red pigment--is not generally stable and may be converted back to myoglobin or metmyoglobin unless it is heated to approximately 130-140F (54.4-60C) yielding nitrosohemachrome. This is also the temperature range at which meat proteins begin to coagulate. Fermentation also coagulates the meat proteins and may stabilize the nitrosomyoglobin.

In addition to their nitrate and nitrite reducing capabilities, the micrococci are known to be proteolytic and lipolytic. They undoubtedly contribute to flavor development by the proteolytic breakdown of the muscle proteins and the production of free fatty acids by lipolysis. During sausage ripening, the concentration of soluble nitrogen compounds increases and may reach up to 25% of the total nitrogen of the meat (Mihalyi and Körmendy, 1967). Part of the liberation of free amino acids has been shown to be due to bacterial proteases. The strong lipolytic activity, characteristic of micrococci, probably results in the liberation of a large variety of fatty acids that are liable to further transformations, leading to the formation of methyl ketones and aldehydes, which may contribute to the unique flavor of these meat products. These reactions are related to the formation of hydroperoxides, from which aldehydes, alcohols, and ketones may be formed (Pezold, 1969).

$$\text{Unsaturated Fatty Acids} \longrightarrow \underset{\text{OOH}}{\text{R-CH-R}_1} \longrightarrow \begin{cases} \text{R-CHO} \\ \underset{\text{OH}}{\text{R-CH-R}_1} \\ \underset{\text{O}}{\overset{}{\text{R-C-R}_1}} \end{cases}$$

Unlike the flavor of cooked meats, which is produced during the heating of precursor substances, the characteristic aroma and flavor of dry sausages is related in part to the hydrolytic and oxidative changes occurring in the lipid portion during the curing process (Cantoni et al., 1967). Specific concentrations of these respective compounds may yield the characteristic flavor while other, higher concentrations may be described as "meat spoilage".

Some researchers have documented the predominance and metabolic activity of the micrococci types (*M. varians*, *M. luteus*, *S. saprophticus*) during the drying phase of some processes (Coretti, 1977). These microorganisms are said to encourage flavor development (via lipolysis and proteolysis), preserve cure color and prevent oxidative rancidity (via catalase activity). In many dry sausages, the product pH increases somewhat as the product dries to completion. This phenomenon is thought to result partially from the proteolytic activity. Micrococcal activity during drying is mainly observed in European-style dry sausages where product pH in the dry room is relatively high at 5.3 - 5.5. These pH values would probably not inhibit these microorganisms. As the product dries, the salt concentration would also increase, favoring the more salt tolerant micrococci.

The genera *Micrococcus*, *Staphylococcus*, and *Planococcus* are now classified under the *Micrococcaceae* family. The microorganisms found in dry sausages have always been regarded as "micrococci" because of the initial classification of these bacteria, and the difficulty of differentiating between the genera *Micrococcus* and *Staphylococcus*. The gram-positive, catalase-positive cocci strains used as starter cultures have also been regarded as "micrococci". However, recent tests on 166 gram-positive, catalase-positive cocci strains in European dry sausages (83) have demonstrated that these organisms are, in fact, predominately (94%) staphylococci (Rheinbaben and Hadlok, 1979). The strains used as starter cultures are also classified as staphylococci. In separate studies, 82 strains isolated from 88 commercial samples demonstrated 61% as staphylococci, 39% micrococci. The staphylococci strains were classified as 20% "*S. xylosus*", 25% "*S. simulans*", and 16% "*S. saprophyticus*" while the micrococci consisted of 36% *M. varians* and 3% "*M. kristinae*" (Fischer and Schliefer, 1980). These findings account for the fermentative activity and acid production that are often observed with some strains and the degree of salt tolerance.

The genus *Staphylococcus* contains three species: *S. aureus*, *S. epidermis*, and *S. saprophyticus*. These species are differentiated mainly by coagulase activity, mannitol utilization, heat-resistant endonucleases, and cell wall content (Table 24). Most strains of *S. epidermis* and *S. saprophyticus* are non-pathogenic while all strains of *S. aureus* are potential pathogens, producing extracellular enzymes and toxins. All three species have been isolated from meat products, and *S. aureus* strains have been associated with food poisoning outbreaks from fermented sausage products (USDA, 1977) which will be discussed later.

S. epidermis and *S. saprophyticus* have optimum growth temperatures in the range of 30-37C (86-99F), some grow at 10C (50F), and some at 45C (113F). *S. epidermis* reduces nitrate and nitrite via reductase enzymes. *S. saprophyticus* will reduce nitrate with the formation of nitrite or ammonia. The metabolism of *S. saprophyticus* is mainly respiratory with slight growth anaerobically in certain media and glucose fermented weakly. *S. epidermis* demonstrates both respiratory and fermentative metabolism. Acid is produced from glucose, usually lactose, and maltose in the presence or absence of air; acid from glycerol and mannitol usually only in air.

Table 24. Characteristics differentiating species of genus *Staphylococcus*[a] (Buchanan and Gibbons, 1974).

	1. S. aureus	2. S. epidermidis[b]	3. S. saprophyticus
Coagulases	+	-	-
Mannitol:			
Acid aerobically	+	d	d
Acid anaerobically	+	-	-
α-Toxin	+	-	-
Heat-resistant endonucleases	+	-	-
Biotin for growth	-	+	NT
Cell wall:			
Ribitol	+	-	+
Glycerol	-	+	d
Protein A	+	-	-
Novobiocin sensitivity[c]	S	S	R

[a] + = most (90% or more) strains positive; - = most (90% or more) strains negative; d = some (less than 90%) strains positive; some negative; NT, not tested.

[b] Four biotypes recognized: see Table 14.4.

[c] R = MIC>2.0 μg/ml; S = MIC<0.6 μg/ml.

Molds and Yeasts

Characteristically, many dry sausages (i.e. Italian salame, Hungarian salami) exhibit mold growth on the product surface, and the respective manufacturers feel these molds are essential to achieve the unique attributes of their product. In addition, the "white" or "grey" surface affords the desirable, "quality" appearance to which the consumer has become accustomed. Since achieving the optimum strain and the degree of mold growth is sometimes difficult, some processors have discontinued the attempt, but they have retained the desirable surface color through the use of white fibrous sausage casings (in Genoa salami). Typically, the mold appears on fresh product during the second to third day of fermentation and it results from the spontaneous contamination from existing product to air, and from the processing equipment. Some manufacturers accelerate this cross-contamination with either fans/blowers or actually rubbing the fresh product with mycelia. These European-style sausages are fermented at 60-75F (15-24C) with 80-90% relative humidity, and subsequently dried at 60-70F (15-21C) with 75-80% relative humidity. In general, drying conditions with less than 75% relative humidity do not afford desirable mold growth. These types of sausages are found primarily in Southern Europe (Table 25) and on the West Coast of the United States (i.e. Italian salame).

Table 25. Approximate utilization of molds (either intentionally or not) in dry sausages (adapted from Leistner, 1972).

Romania	100%	East Germany	1
Italy	95	West Germany	1
Bulgaria	90	Poland	1
Hungary	80	Russia	0
Switzerland	70	Finland	0
Spain	50	Holland	0
Austria	30	Norway	0
France	5	Sweden	0
Belgium	5	Denmark	0
Yugoslavia	5	Great Britain	0

In the aging process, molds have two important functions:

1. The mycelial coat on the product surface has a tendency to regulate moisture loss. Within limits, this natural control mechanism may compensate for changes in the relative humidity of the environment, achieving a more uniform drying rate from the sausage center.

2. As the mold develops and uniformly covers the surface, enzymatic action on the fats and proteins can influence the flavor and aroma that is distinctive of the product. Large drying chambers can have a pronounced "ammonia aroma" resulting from proteolytic action on meat proteins.

In addition, since all molds have catalase activity and form a coating over the surface, they most likely reduce any tendency toward rancidity development by preventing oxygen penetration into the meat. Many molds also reduce nitrates to nitrite, enhancing the surface color.

Many processors, with less sophisticated environmental controls, utilize the mold growth as an "indication" of the proper drying conditions (Figure 21). Excessive humidity can result in undesirable "green" or "black molds". This may result from both sporulation of the desirable molds and/or selection of different species. "Ideal" conditions in Italian salame manufacture result in a uniform covering of white or blue-white conidia and mycelia.

Figure 21. Many processors use the mold growth as an indication of proper drying conditions (photo courtesy of San Francisco Sausage Co.).

In West Germany, two species of molds have been approved (as of 1978) as starter cultures. The "white molds", Penicillium candidum, and the "blue molds", Penicillium roqueforti are both atoxic, develop well, and yield good results. Both these species also form part of the microbial flora of Hungarian and Italian salamis. Another mold, P. nalgiovensis, has also been documented as a desirable starter culture and is included in the group, P. canescens (Leistner, 1972). Several strains of Actinomycetes and Streptomyces have also been described for use in dry sausages (Coretti, 1977) and Thannidium elegons has been proposed in the United States for use to "rapidly age" fresh meats (Anonymous, 1978).

The species P. candidum, P. roqueforti, and P. nalgiovensis have specific lipolytic enzymes that degrade fat, producing a strong and "slightly irritating" flavor that is characteristic of Hungarian and Romanian salamis. These molds also synthesize proteolytic enzymes and amylases which yield products that influence the flavor and aroma. Many molds also produce antibiotics which may affect the bacterial microflora, and as a result, any contributing factors of their metabolism. None of these three species produce the enzyme cellulase which would preclude the use of fibrous casings. Some "wild" molds growing on the fibrous casings will degrade the material and result in sausages on the floor of the drying room!

Certain prerequisites are desirable in the selection of a specific mold strain for use as a meat starter culture (Schiffner et al., 1978).

1. The species should not be toxic. Although the predominant molds isolated from dry, European-style sausages are penicillin or penicillia and Scopulariopsis, some aspergilli may also be present. A strain of Aspergillus flavus isolated from an Italian-type product was found capable of aflatoxin production under certain conditions. Subsequent studies with "inoculated salamis" demonstrated that low temperatures, low humidities, and smoking prevented any aflatoxin production.
2. The conidia and mycelia should exhibit a color that ranges between "white-grey" or "grey".
3. They must cover the casing uniformly.
4. The mold must germinate and grow rapidly.
5. The mold must develop a specific and consistent flavor.
6. The mold should not possess significant cellulase activity.

P. candidum forms a mycelial mat of pure white conidia and grey-white mycelia. It grows well and the germination phase requires 3 to 4 days in typical European processes.

P. roqueforti forms a mycelial mat of blue conidia and dark grey mycelia. It grows very well and the germination phase lasts 2-3 days.

P. nalgiovensis grows well within 3 days and yields a heavy "cover" which is grey-white.

Although many processors continually attempt to produce their product with a uniform mold covering, many sausages in the market lack uniformity. As of 1978, in Germany, only 1% of the sausages utilize mold starters, which is well below the demand for these type of sausages. The isolation and cultivation of molds from imported sausages has not generally resulted in the successful implementation of these as starters in sausage-making. The majority of these sausages contain a mixed mycotic flora that has proliferated under a specific set of conditions. Isolated strains may not develop simultaneously in another formulation and process.

From Hungarian salami, all the following species of molds have been isolated (Leistner, 1972):

Scopulariopsis brevicaulis

Penicillium camemberti

P. comune

Scop. alboflavescens

P. expansum

P. myczinski

P. simplicissimum

P. candidum

Commercial mold starter cultures consist of spore suspensions of 10^6 - 10^8 spores per milliliter in a nutrient medium. The shelf-life is approximately 30 days. Various methods of inoculation range from spraying, immersion, and/or adding directly to the sausage mix. Generally, soaking the sausages (1/2 hour, warm water) is the most sanitary and cost effective. The nutritive "carrier" medium is also deposited on the sausage surface and will provide initial nourishment to enhance germination and growth.

Generally, high humidity is required to effect germination. The following processing parameters are recommended (Leistner, 1972).

P. nalgiovensis

40 hours @ 22C (72F) 90% RH

20 hours @ 20C (68F) 85% RH

10 hours @ 18C (64.4F) 80% RH

P. candidum/P. roqueforti

2-3 days @ 18C (64.4F) 95-98% RH

2-3 days @ 15C (59F) 90% RH

The respective cultures do not reach a stationary phase until long after this initial germination. The mycelial development is less sensitive to the environmental conditions and can proliferate during typical, subsequent drying processes.

In many traditional processes, the bacterial fermentation and the establishment of the mold generally occur simultaneously via random contamination. Since the mold germination and growth is somewhat slower, the bacterial action can sometimes precede the mold growth depending on the respective inoculum levels. The relatively high temperatures and humidities required for spore germination are not always desirable for the development of the proper lactic microflora in European processes, especially when starter cultures are not employed. Consequently, many processors may initiate the fermentation process in advance at lower temperatures and inoculate the mold spores by immersion of the sausages after 24 hours.

Molds used in commercial practice can be attacked by mutants, yielding a weak mold with the absence of the characteristic sausage aroma and flavor. In addition, mold viruses can invade the culture inhibiting its growth.

Some European researchers have also documented the utilization of certain yeasts, particularly of the Debaryomyces Family, in the manufacture of dry sausages (Coretti, 1977). The yeasts accelerate the cure-color development and result in a distinctive aroma. Some of the yeast strains have been experimentally employed in combination with lactic bacteria and micrococci, but commercial yeast starters are generally not yet available.

CHAPTER 4

Factors Affecting Meat Fermentations

Introduction

As the technology of meat fermentation has increased, the responsible lactic acid microorganisms have been isolated and reintroduced as pure starter cultures for the manufacture of these fermented meat products. Through the expanded use of starter cultures, the initial type and number of microorganisms inoculated into a specific meat product has become more consistent. As a result, other environmental factors affecting the meat fermentation have become more apparent. Although these formulation and process parameters affect both "wild" and "cultured" fermentations, the constant inoculum afforded by the use of a starter culture has allowed the other environmental factors to be more readily observed. The purpose of this chapter is to summarize these environmental factors.

The rate of fermentation and the ultimate pH of the meat product are directly influenced by the specific formulation and processing conditions, as well as the type and "activity" of the culture employed. Since the safety and quality of the product are dependent on the rate and extent of acid production, a thorough understanding of these environmental parameters is essential to the total control of the product.

Formulation

Meat

Although the meat comprises the majority of a sausage formulation, it is often overlooked as a factor affecting fermentation. The lactic microorganisms operate in the water phase of the formulation, thus any factor affecting the amount of available water will influence the microbial activity. The greater the moisture content of the meat, the greater the fermentation rate. The fat content is also important as it is inversely proportional to the lean meat percentage. In general, the higher the lean percentage (i.e. lower fat), the greater the moisture, resulting in a more rapid pH drop. The contribution of glycogen from lean tissue appears to also have an effect on total acidity (Acton et al., 1975). The use of freeze-dried meat as an ingredient in dry sausage will retard the initial fermentation rate due to the reduced moisture (Klettner, 1980).

The buffering capacity (i.e. the ability of the meat to absorb acid) will also affect the rate of pH drop. The greater the buffering capacity, the more acid that must be produced by the bacteria prior to lowering product pH. This results in a slower overall fermentation time. The initial meat pH is also important as to subsequent fermentation time and final product pH. Meats with higher pH values will require more acid production to achieve the same end point (Acton et al., 1977).

The ratio of fresh vs. frozen meat will influence the fermentation in several ways. Since microorganisms have an optimum growth temperature, the internal product temperature is critical to the fermentation time. A formulation high in frozen materials will take longer to achieve the desired fermentation temperature, thereby extending the total incubation period. In addition, frozen meats are often somewhat dehydrated and exhibit rapid drip loss during thawing. A reduced moisture level, attributed to previously frozen raw materials, will reduce the rate of pH drop.

Many processors note that pork and beef products ferment at a faster rate than their all-beef counterparts. Several hypotheses include higher lactic contamination levels found in pork and/or the higher thiamine concentrations. In addition, beef will generally have a higher initial pH and greater buffering capacity.

The microbial flora of the raw meat materials, attributed to prior handling conditions and sanitation, may also influence subsequent fermentation through either direct microbial interactions and/or indirect chemical changes in the raw materials. High levels of lactic acid microorganisms may actually increase the rate of fermentation while undesirable microorganisms (pseudomonas, yeast, etc.) may produce end-products, either prior to, or during the lactic fermentation, that affect the flavor. High initial numbers of certain yeasts can compete successfully with the lactic acid microorganisms and retard pH drop through the production of more basic end-products, including alcohol. Since most yeasts are acid tolerant, this can also occur after the fermentation has been completed, resulting in a pH increase and the development of off-flavors during drying. A high oxygen content in the meat may induce a microbial oxidation of a portion of the available carbohydrate with the production of carbon dioxide, water, alcohols, and carbonyl compounds (DeKetelaere et al., 1974). Micrococci present in the early stages of fermentation can also be responsible for the complete conversion of carbohydrate. Either of these situations can result in a higher pH than expected. Many of the flavor differences and flavor problems in fermented sausages can be attributed to the differences in how the added carbohydrates are metabolized by various groups of bacteria that predominate in a unique process (Acton, 1977).

Non-Meat Ingredients

Salt is added to fermented meat products to achieve the desired bind (i.e. myosin extraction), flavor, and preservative qualities. Salt is the major component that allows the lactic acid bacteria to predominate and inhibits many undesirable microorganisms. In general, fermented sausages are formulated with 2.0 to 3.5% salt, depending on the nature of the product. Although the lactic micro-organisms responsible for the fermentation are salt tolerant, the brine concentration (% salt/% moisture) of the formulation directly affects their performance. A salt level of 2% is regarded as a minimum to achieve the desired bind, and no major differences in fermentation rate are generally observed up to 3% (Zaika et al., 1978). However, salt concentrations, in excess of 3% (i.e. 5-6% brine) will begin to lengthen fermentation time (Figure 22). Since salt is commonly employed as a "carrier" for curing agents and flavorings, the calculated salt content of the formulation should also include this portion.

Although salt definitely contributes to the stability of fermented meats, a reduced pH is also essential. An excess of salt in the initial formulation will not compensate for an extended fermentation time. Food poisoning staphylococci, which are a primary concern in fermented products, are more salt resistant than most lactic acid bacteria and, therefore, can better tolerate increasing salt con-centrations. A sausage formulation containing 4% salt may actually favor the growth of any contaminating staphylococci over the lactic acid microorganisms.

Various sugars (dextrose, sucrose, corn syrups) are added to the formulation to achieve the desired flavor, texture, and yield char-acteristics. They also provide the necessary fermentation substrates for the production of lactic acid (Figure 23). The type and amount of sugar can directly affect the ultimate pH of the product (Figure 24). Simple sugars, such as dextrose, are readily utilized by all lactic acid bacteria and their availability tends to be the limiting factor for sausage fermentation. It has been noted by some researchers that 1% fermentable carbohydrate concentration will generally effect a decrease of 1 pH unit. Where the initial pH is 6.0, 1% dextrose should be added to promote an adequate drop in pH (Acton et al., 1977). If the fermentation is allowed to continue to completion, the final pH is directly proportional to the initial dextrose level, up to approximately 0.5% (Figure 25). It has been generally recommended that sausage mixtures contain a minimum of 0.75% dextrose, to afford a surplus of a known, fermentable carbohydrate. Cane sugar, or sucrose, also can be utilized by most lactic bacteria and will generally result in a "less acid tasting" product when compared to dextrose at an equivalent pH. This probably results from either the greater "sweetness" from the residual sugar, or the utilization of either the fructose/dextrose component to yield a more basic end-product (i.e. dextran, levan).

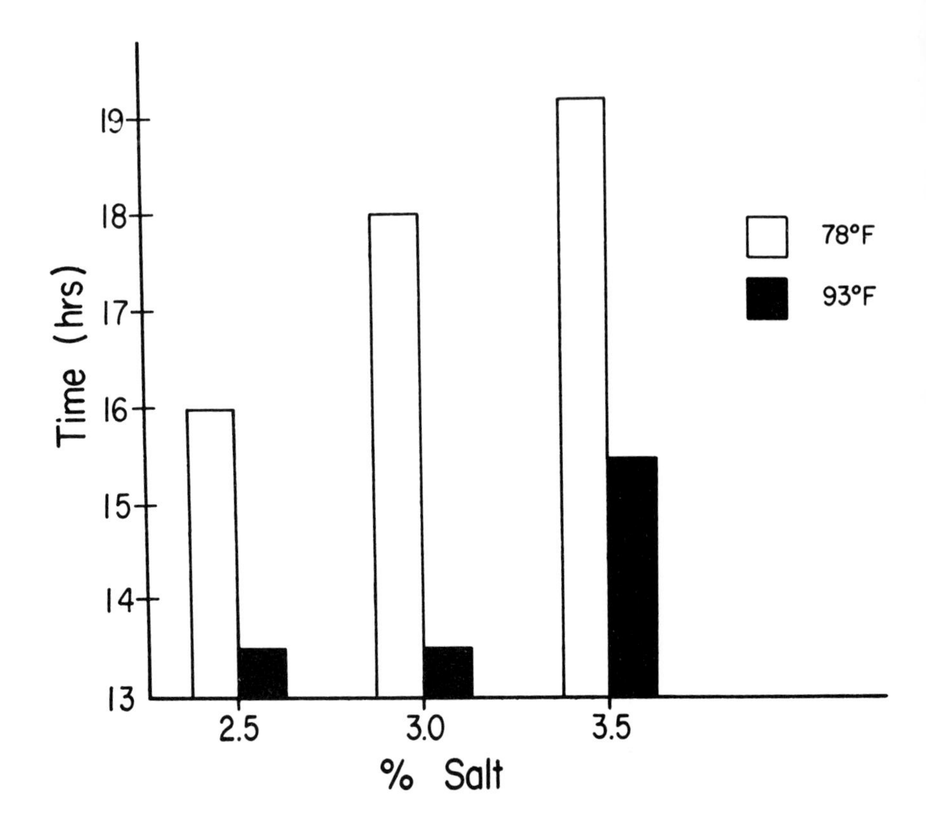

Figure 22. The effects of salt concentration and temperature on the sausage fermentation time of a lactobacilli blend to reach a pH 5.1.

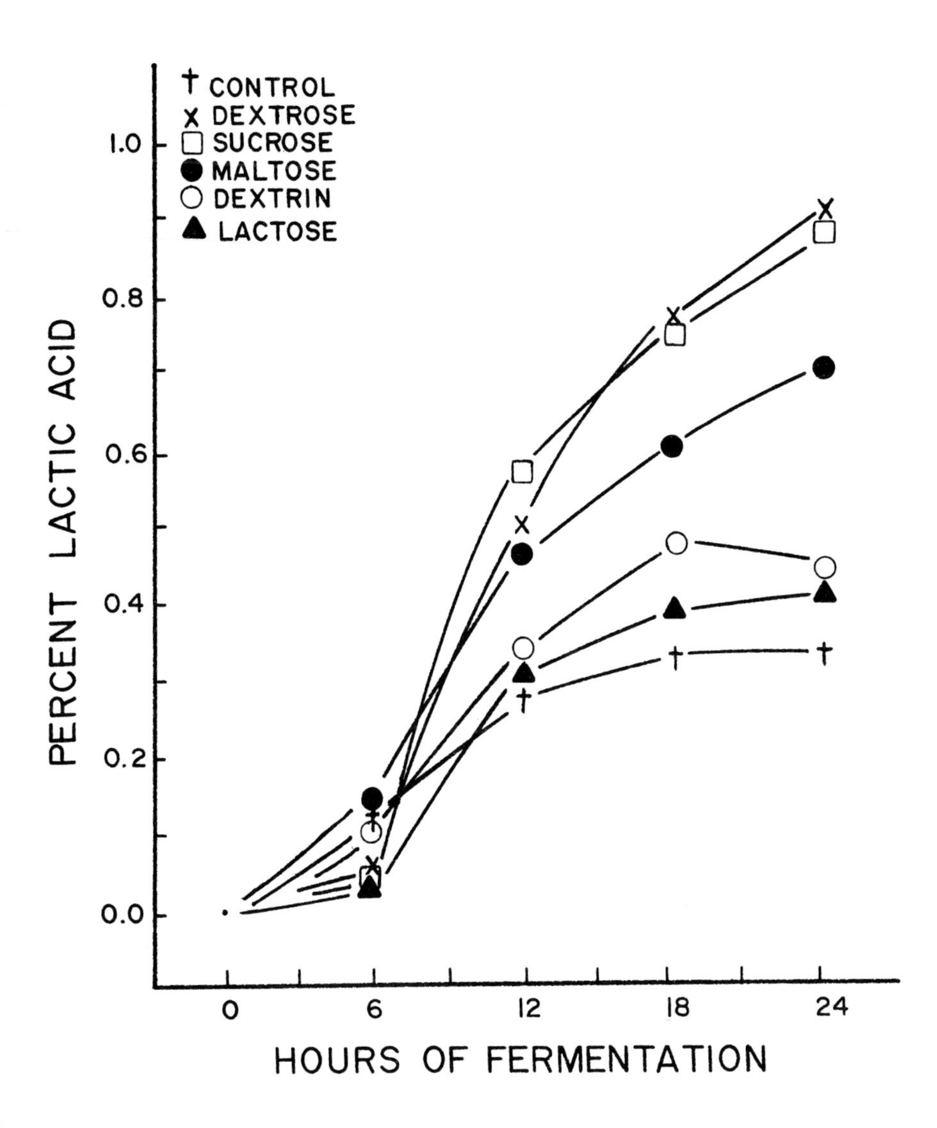

Figure 23. Rate of lactic acid accumulation in fermenting sausage containing 0% or 1% of various carbohydrates (from Acton et al., 1977).

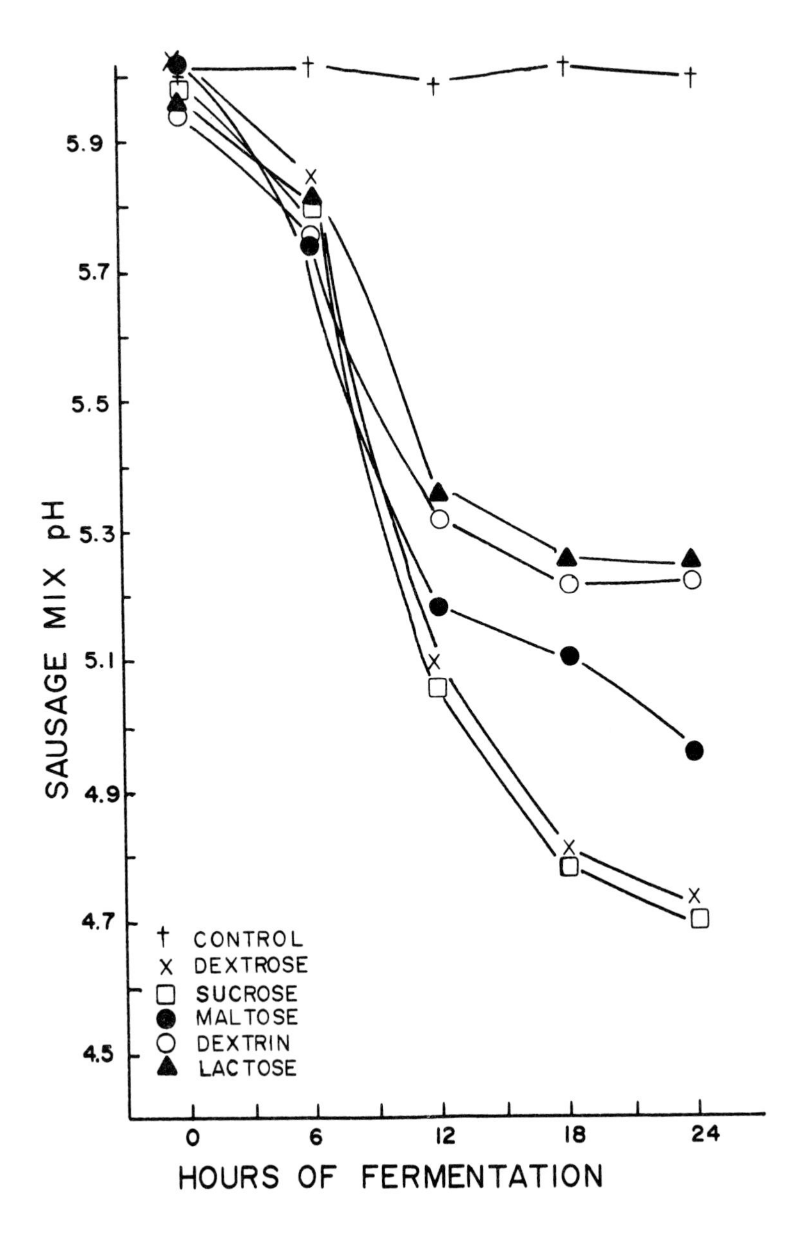

Figure 24. Rate of pH reduction in fermenting sausage containing 0% or 1% of various carbohydrates (from Acton et al., 1977).

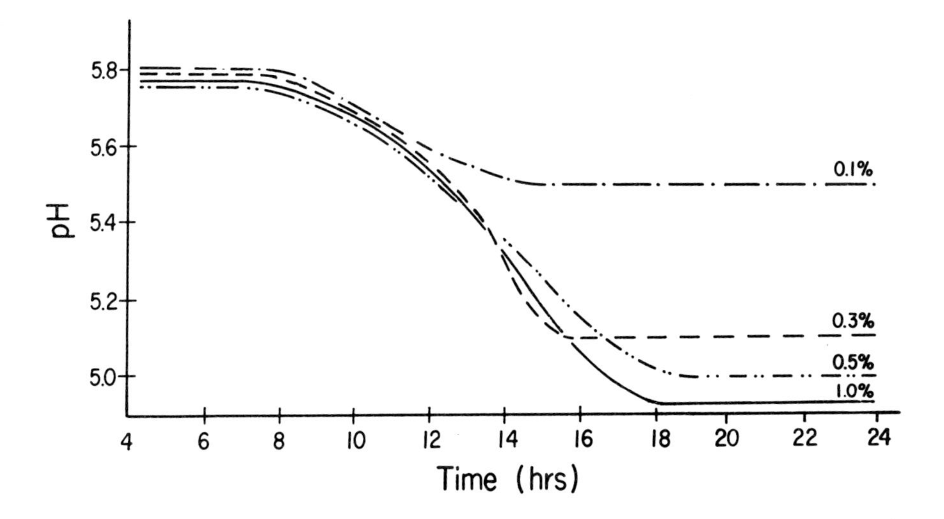

Figure 25. The effects of varying initial dextrose concentration on sausage fermentation by a lactobacilli blend.

More complex carbohydrates such as corn syrup, dextrin, flour, and starches also can be fermented to various degrees depending upon their availability and the specific culture employed. These carbohydrates ferment more slowly and are not of practical significance where simple sugars are present, except in some dry sausages where unique flavors are produced over extended drying periods. The amount of acidity obtained from corn syrups was found dependent on the quantity of simple carbohydrates, dextrose and maltose, initially available (i.e. dextrose equivalent) (Acton et al., 1977).

Excess concentrations of added carbohydrate (>2%) can also reduce the rate of fermentation through the binding of the available water.

Certain natural spices, typically used in the formulation, can have a direct effect on the rate of fermentation by stimulating acid production in the bacteria (Figure 26). Often, the same processor with various products differing in spice composition observes differing rates of fermentation. Generally, this stimulation is not accompanied by an increase in the bacterial population and does not result from any contaminating microflora in the spices. Black pepper, white pepper, mustard, garlic powder, allspice, nutmeg, ginger, mace, cinnamon, and red pepper have all been shown to stimulate acid production to varying degrees depending on the concentration and the culture employed. Generally, lactobacilli are more stimulated than the pediococci strains. Combinations of certain spices can often yield a shorter fermentation time than individual spices alone (Table 26, Figure 27). The degree of acid stimulation can also be dependent on the origin and type of the particular spice component (i.e. Lampong black pepper vs. Brazil black pepper). Recent studies have identified manganese as the factor in spices responsible for the enhancement of the acid production (Zaika and Kissinger, 1982). Stimulatory activity of acid extracts of spices increased with increasing manganese concentrations. Fermented sausages without spices but with added manganese (10^{-5} M) developed a similar level of activity to those sausages with added spices. This finding correlates to similar results documented in an early U.S. patent covering meat starter cultures (Chaiet, 1960). In addition to an increased rate of acid production, the ultimate pH of the product is often lower when either natural spices or manganese is added to the sausage mix.

Inhibiting effects also have been observed with some natural spices and, particularly with the extracts of some spices (i.e. pepper), generally the volatile fraction. Some results have suggested that the normal levels of pepper used in sausages are not adequate to inhibit bacterial growth unless it is used as the essential oil (Salzer et al., 1977). This probably accounts for the somewhat longer fermentation times generally observed with liquid spice blends compared to their natural spice equivalents. In view of recent findings, if this is a true inhibition, or just a lack of stimulation by the manganese content of the natural spices, remains to be seen.

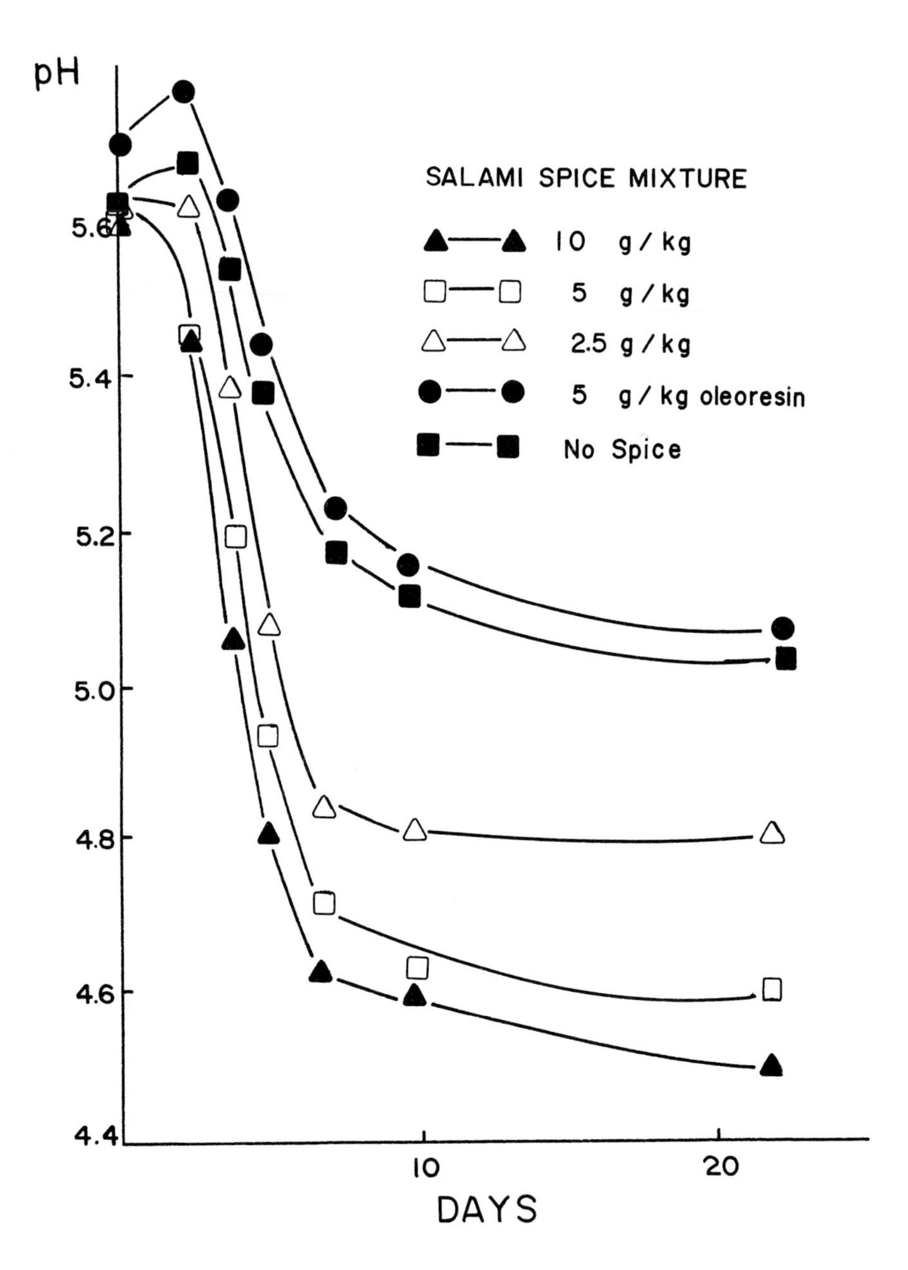

Figure 26. pH-course of dry sausage during fermentation in the presence of varying spice mixtures (from Ingolf and Skjelkvale, 1982).

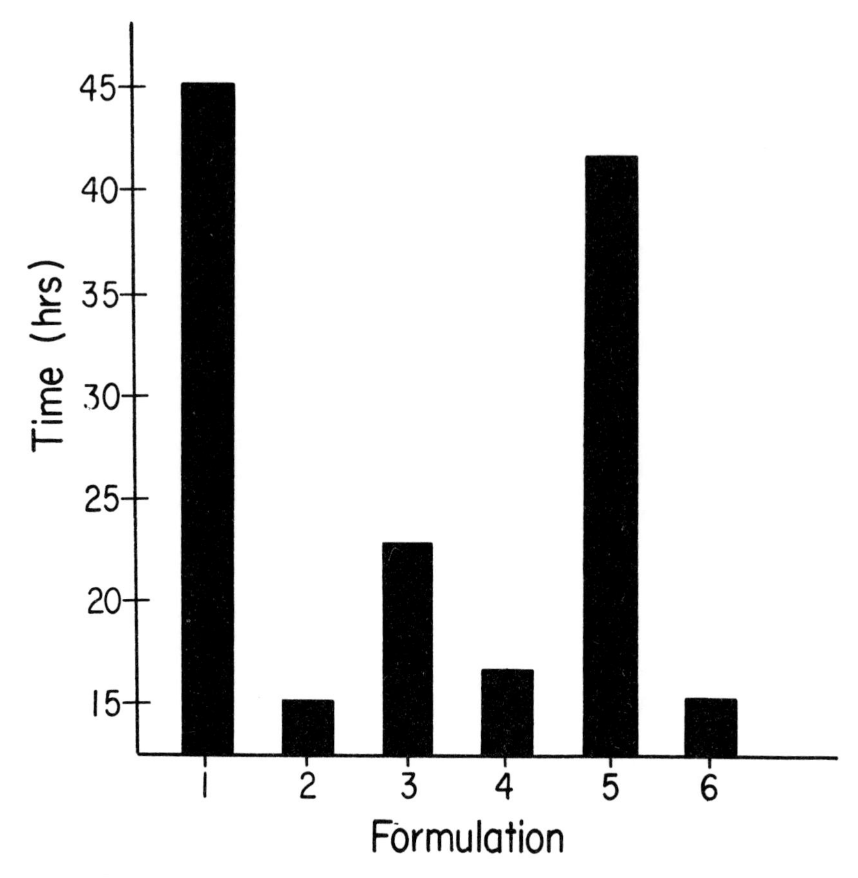

Figure 27. The effect of spice formulation (Table 26) on the sausage fermentation time of a lactobacilli blend to reach pH 5.0.

Table 26. Culture stimulation by spices in fermented sausage.

Ingredients	Formulation (%)					
	1	2	3	4	5	6
Lean Beef (90/10)	96.0	93.8	95.8	95.8	96.0	95.6
Salt	3.0	3.0	3.0	3.0	3.0	3.0
Dextrose	1.0	1.0	1.0	1.0	1.0	1.0
Sodium Nitrite	0.015	0.015	0.015	0.015	0.015	0.015
Corn Syrup Solids	-	1.5	-	-	-	-
Ground White Pepper	-	0.17	0.17	-	-	0.17
Ground Black Pepper	-	0.23	-	0.23	-	0.23
Garlic Powder	-	0.013	-	-	0.013	0.013
Monosodium Glutamate	-	0.2	-	-	-	-
Sodium Erythorbate	-	0.04	-	-	-	-
	100.0	100.0	100.0	100.0	100.0	100.0

Other non-meat ingredients typically encountered in fermented sausage formulations can exert different effects on the microbial fermentation. The practical significance attributed to each component is dependent on the concentration, the culture activity, the remaining formulation and the processing conditions. Generally, liquid smoke flavorings and antioxidants retard fermentation rate (Donnelly et al., 1982; Raccach, 1981). Phosphates, depending on the type and amount, will function as a buffer, increasing the initial pH and the "lag time" prior to observing a pH decrease (Figure 28). Milk powder, soy proteins, and other dry powders can slow the fermentation time by binding the available water. Formulations with added water will ferment at a faster rate. Sausages containing sodium nitrite will generally ferment slower than those without nitrite, but the degree of difference is largely dependent on the specific culture strain (Sutic, 1978).

Process

The processing conditions affect both the rate of fermentation and the ultimate pH of the product. In most "wild" fermentations, the processing conditions select the type of microorganisms that will predominate which, in turn, determine the characteristics of the fermentation and the final flavor of the product. Processing parameters such as time, temperature, humidity, and smoke can be utilized effectively to control the total process and, thus, prevent problems.

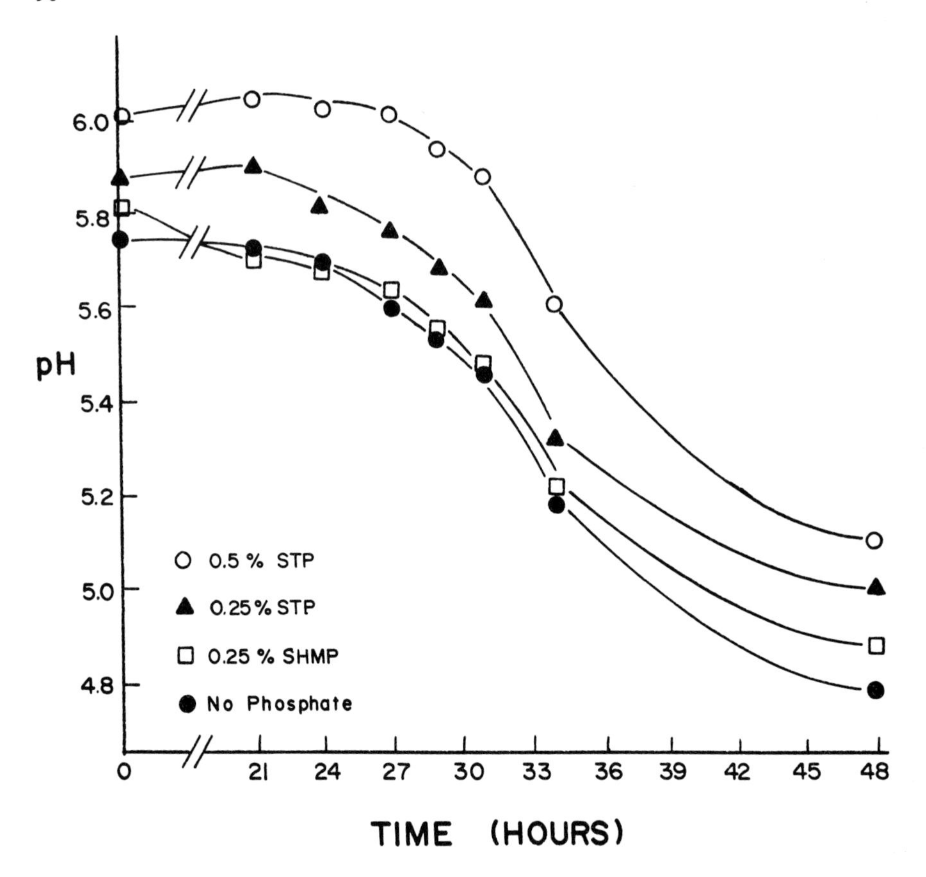

Figure 28. Effect of phosphates on fermentation of salami (1% dextrose) at 75F (23.9C)

STP = sodium tripolyphosphate
SHMP = sodium hexametaphosphate

Product temperature largely will determine the metabolic activity of the microorganisms present (Figure 29). The lactobacilli and pediococci employed as starter cultures have optimum growth temperatures of 89.6F (32C) and 98.6F (37C), respectively. Their initial performance in lowering pH will be dictated by the product temperature and the time held at that temperature. Slower fermentation rates will be observed as the product temperature deviates from the optimum (Figure 30). In many instances, slower fermentations at lower temperatures are more desirable in controlling ultimate pH and developing flavor, color, and other product characteristics. Lower temperatures are also utilized to control undesirable, pathogenic microorganisms. Higher finishing temperatures are employed in many products to stop the fermentation phase at a desired pH. Although internal product temperatures of 145-155F (63C-68C) usually are required to effectively kill the lactic microorganisms, temperatures of 115F to 125F (46C to 52C) for extended periods can be employed to half the acid production for lactobacilli and pediococci, respectively. An important factor to be considered in accounting for fermentation time is the "come up" time to achieve the desired product temperature. This is influenced directly by the initial product temperature, the fermentation temperature, the load in the house, air circulation, and humidity. Depending upon the degree of internal control, changing weather conditions may affect fermentation times by generally lowering overall temperatures and humidity. Some processors observe seasonal variations in fermentation time with basically the same formulation and process.

Higher humidity favors more rapid fermentation since any drying of the product reduces available water. Dry sausages held with static air conditions usually ferment faster, and subsequently dry better, than those fermented with high air circulation. Although the latter conditions will effect more rapid heating, the surface of the meat will tend to dry out--irreversibly sealing the surface pores and preventing uniform drying.

Smoke is definitely inhibitory to microorganisms. In larger diameter products, smoke will not exert a major effect on fermentation since the smoke penetration is limited to the surface. However, in smaller diameter products (13-18 mm), heavy smoke application at the beginning of the fermentation cycle may retard the microbial activity. Smoke may also have a tendency to dry the product.

Casing diameter is also important in predicting fermentation time and controlling ultimate pH. Larger diameter product will generally exhibit lower ultimate pH's than the identical formulation processed in smaller diameter casings (Figure 31). Even though the fermentation may initially proceed slower in the larger diameter product, due to slower heat penetration, the fermentation is also more difficult to stop with subsequent heat treatment or drying.

In dry sausages, the ultimate pH has been attributed mainly to the concentrations of ammonia and lactic acid and the buffering capacity of the proteins. A significant negative correlation was found between

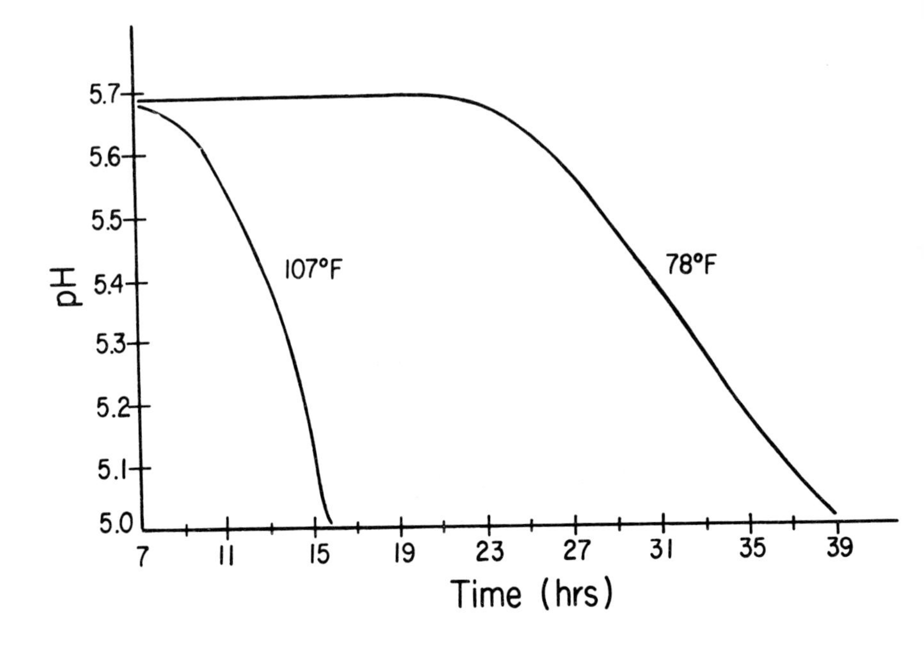

Figure 29. The effect of temperature on sausage fermentation by a pediococcus strain.

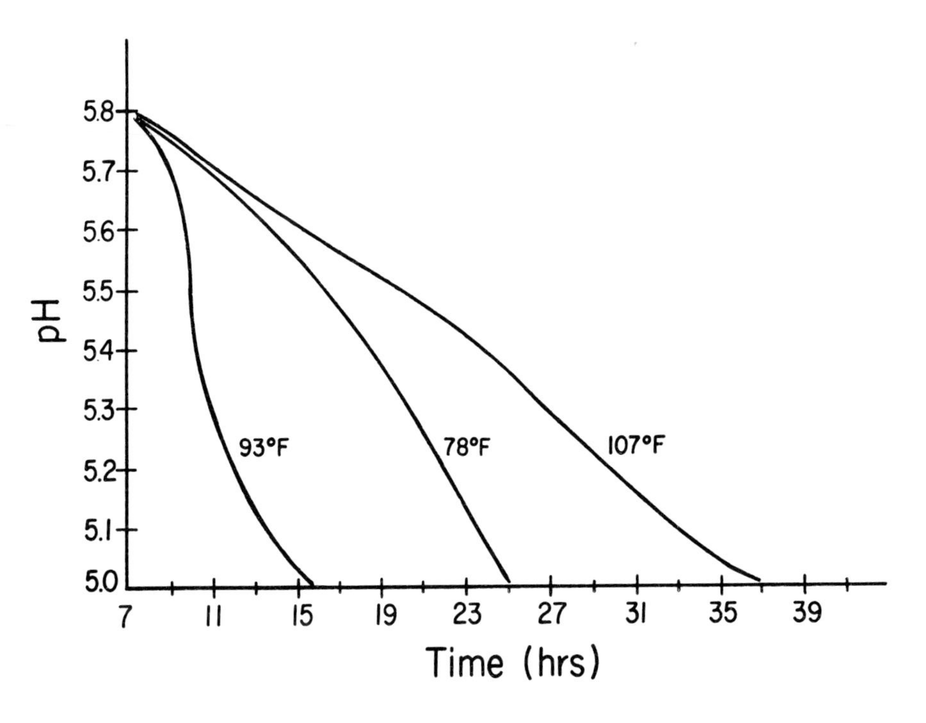

Figure 30. The effect of temperature on sausage fermentation by a lactobacilli blend.

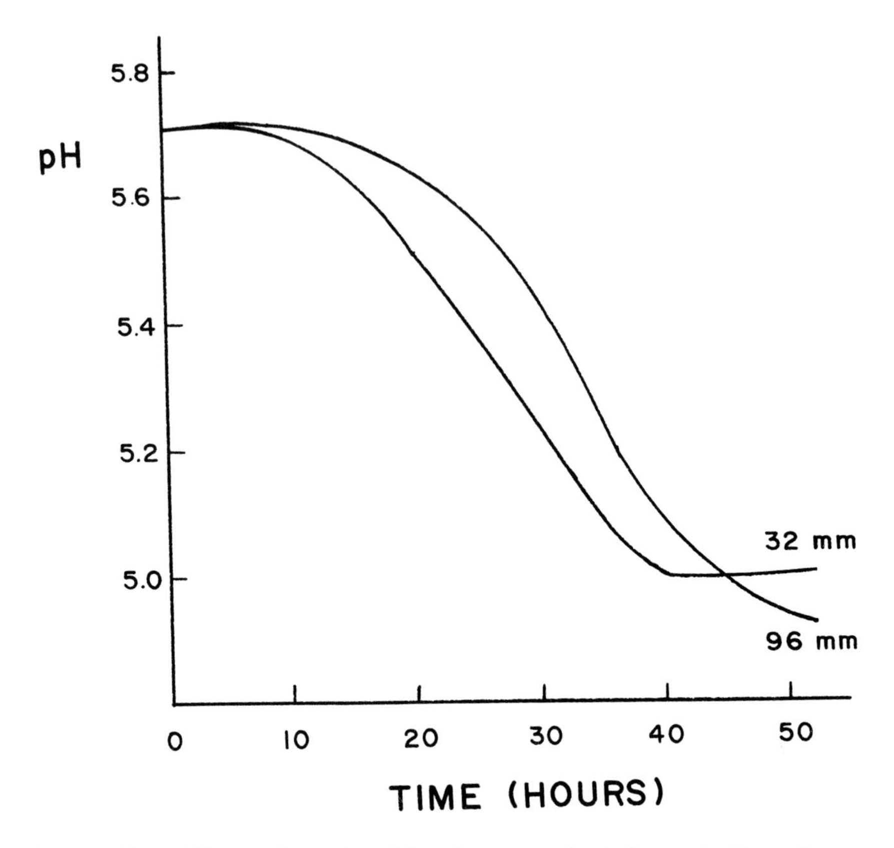

Figure 31. Effect of casing diameter on salami fermentation at 75F (23.9C).

moisture content and pH (Demeyer et al., 1978), possibly related to the lesser degree of lactic acid dissociation. This may explain the gradual increase in pH in many dry sausages during the aging process, and the increase in pH of some semi-dry products when fully cooked.

Formulation and process uniformity determines fermentation consistency. Uniform distribution of formulation ingredients is essential to maintain a constant micro-environment within each batch, and from batch to batch. Non-uniform distribution of salt, cure, sugar, spices, and/or starter culture may yield variations in the rate of fermentation and ultimate pH from sausage to sausage. This can result in flavor, texture, and stability problems in those pieces not adequately fermented. Variation can also result from product "positioning" within the smokehouse and dry room. Areas of higher temperature and humidity will generally ferment faster to a lower pH.

Although not a common occurrence, process contaminants may retard fermentation rate. Boiler treatment compounds, when improperly used, can be toxic to the lactic microorganisms when exposed through a high humidity process. Sanitizing agents definitely are lethal to starter cultures and drastically can reduce culture activity through direct contact. Containers used to dilute and distribute the starter culture should be rinsed thoroughly after sanitizing. Processing equipment should also be rinsed so as to eliminate residual sanitizers prior to formulation.

Culture

The type and activity of the microbial starter culture or "seed" inoculum is also a key element in structuring a consistent fermentation process. Knowledge of the optimum growth temperature and the effects of various additives will provide insight into any product variation and problems. Optimum storage and handling conditions for the culture are critical in maintaining consistent performance, as is uniform distribution of the culture throughout the batch. The distribution medium, usually tap water, should be periodically monitored as to excessive levels of chlorine or metals, if culture activity appears to decrease. The culture should never be directly mixed with the cure, salt, or other dry ingredients prior to blending into the meat.

Meat culture "activity" generally refers to the relative ability of the starter to produce lactic acid in a designated meat system. The rate of pH decline in meat increases with increasing culture "activity". Although the total number of microorganisms definitely influences the fermentation rate, the activity, or acid-producing ability, of each cell can be affected by the growth medium, the harvest conditions, the method of preservation prior to subsequent inoculation into the meat, and final meat product formulation and processing parameters.

A commercial culture medium for P. cerevisiae (acidilactici) has been described as corn steep water, 25 lbs.; powdered non-fat dry milk, 50 lbs.; dextrose, 100 lbs.; yeast autolysate, 13 lbs. 5 ozs.; potassium dihydrogen phosphate, 14 lbs. 4 ozs.; disodium hydrogen phosphate, 9 lbs. 9 ozs.; and water to bring the mixture to 600 gallons (Rothchild and Olsen, 1971). The culture medium is inoculated with P. cerevisiae (acidilactici) in a conventional manner and is incubated for 8-10 hours at a temperature of 89.6-98.6F (32-37C). The bacteria are then separated from the medium (usually by centrifugation), concentrated, mixed with other stabilizing and nutrient materials, and frozen as rapidly as possible. Recommended storage temperature is -5F (-20.6C) or below (optimum -15F, -26.1C).

Although each commercial culture manufacturer has proprietary media formulations and processing conditions to enhance "activity", it generally has been found that a meat culture must be grown in the presence of sodium chloride (minimum 0.5%) as to retain its salt tolerance. Acid-producing capability and salt tolerance are regarded as the two primary attributes of a Pediococcus meat starter culture.

Other media components, such as various mineral salts, stimulate growth, acid-production, and/or subsequent "activity". In particular, manganese salts have proven very effective to increase the "activity" of both P. acidilactici and P. pentosaceus strains employed as meat starters (Raccach, 1981).

Initially, a strain of P. cerevisiae (acidilactici) was chosen as an appropriate meat starter since it readily survived lyophilization which was the method chosen at the time to preserve and distribute these cultures (Deibel et al., 1961). However, the "lag phase" re-required to rehydrate these cultures later proved insufficient, and frozen concentrates were introduced. Frozen cultures are presently the predominant form being utilized in the United States. Freeze-dried meat cultures still are widely employed throughout the rest of the world, presumably due to problems encountered in distribution of frozen products and the relatively slower, fermentation times required

Culture "activity" must sufficiently be retained during storage to achieve the necessary performance in the meat system. Various stabilizing agents such as glycerol, non-fat dry milk, monosodium glutamate, malt extract, alkali metal glycerophosphates, glutamic acid, cystine and/or dextran often are combined with the culture concentrate in a freezing medium as to afford protection and retain "activity" (Rothchild and Olsen, 1971). Culture concentrates usually exhibit 10^9 to 10^{11} cells per ml (g), and liquid nitrogen or carbon dioxide are employed to freeze the culture as rapidly as possible.

An alternative commercial method to preserve and distribute meat cultures employs a liquid anti-freeze agent (Storrs, 1980). A conventional cell paste is diluted with one or more water freezing point depressants which are water soluble, non-injurious to the bacteria, and do not form crystals when cooled to -40F (-40C). Such depressants include polyhydric alcohols, sugars, and other water-soluble inorganic and organic compounds used as cryoprotective agents. The process reduces the water activity of the medium by utilizing a minimum of 40-50% of the depressant and provides a non-frozen concentrate that can be cooled to -40F (-40C) to preserve viability and "activity". The liquid form enhances the handling characteristics for subsequent sampling, and the medium inhibits the formation of damaging ice crystals. In addition, the culture concentrate can be warmed during normal distribution without as much loss in viability and "activity" as that encountered with frozen cultures that are thawed.

Meat fermentation is a biological acidulation whereby unique product characteristics result from a microbial lowering of meat pH. It cannot be emphasized too strongly that the biological nature of the process makes it susceptible to many environmental factors, which must be controlled to yield a consistent product. The use of a commercial starter culture provides an additional control mechanism through a consistent inoculum; however, it will not replace a thorough knowledge and control of the total process. Both "cultured" and "wild" fermentations involve similar lactic acid microorganisms whose function must be properly understood to achieve the desired result.

Based on scientific research and practical experience, a "trouble shooting" guide has been compiled to identify possible, causative factors when inconsistent fermentations and products are observed although a starter culture is being employed (Table 27).

Table 27. Troubleshooting guide--fermented meats.

Problem	Possible cause
slow fermentation	-frozen culture allowed to thaw and subsequently held too long before dispensing into meat. Microorganisms exhaust nutrients in can, reduce pH, lowering culture "activity"
	-environmental temperatures/humidities during fermentation inconsistent with recommended culture optimum
	-secondary growth in meat of contaminant microorganism producing end-product which buffers pH drop
	-laydown procedure at cold temperatures resulting in extended lag phase at beginning of fermentation cycle

Table 27 (Continued)

Problem	Possible cause
	-cheese in product may contain phosphate which buffers pH drop; also has tendency to absorb moisture from surrounding meat
	-sausage entering smokehouse colder than normal, i.e. using frozen meat
	-spice formulation adjustment that either decreases acid stimulation or inhibits culture
	-excessive salt or cure addition
	-culture contact directly with curing components
	-high fat formulation, reducing moisture content
	-larger diameter product, slower heat transfer
	-rapid moisture loss in product
fast fermentation	-temperature/humidity higher than normal
	-spice formulation adjustment favoring culture
	-excessive water addition
	-product delayed prior to entering house, resulting in higher initial temperature/more time
	-leaner product, i.e. more moisture
	-pork containing product
	-smaller diameter product processed at high humidity
	-initial meat pH lower than normal
inconsistent fermentation	-inadequate culture distribution, resulting in "hot" and "cold" spots in meat mixture
	-inadequate distribution of salt, cure, spices, dextrose
	-laydown procedure that causes some "batter" to dry out
	-various initial product temperatures

Table 27 (Continued)

Problem	Possible cause
	-"laid down" product and directly processed product in same smokehouse; culture "acclimation" in laid down product resulting in faster fermentation
	-products with different spice formulations, meat components, casing diameters
	-uneven temperature/humidity in fermentation chamber
	-uneven humidity in dry room causing different drying rates
too low pH	-failure to monitor fermentation
	-excessive carbohydrate source
	-insufficient heat processing to retard fermentation
no fermentation	-culture not added
	-culture inactivated by direct contact with salt, cure components, or heavily chlorinated diluent water
	-non-compliance with recommended handling temperatures after thawing
	-insufficient carbohydrate added to sausage mixture
	-excessive salt content
	-antibacterial agents added to meat mixture (i.e. preservatives, chemical boiler treatments via steam, antibiotics in meat)
"mushy" product	-over-working at mixer, chopper or grinder
	-excessive fat content
	-insufficient salt level or no salt added
	-spoiled raw materials
	-proteolytic microbial contaminant
insufficient moisture loss	-excessive humidity

Table 27 (Continued)

Problem	Possible cause
	-excessive air speed and/or too low a humidity, "sealing" surface pores. No moisture migration from product
	-excessive smoke initially that coagulates surface proteins, retarding moisture migration
souring of product, post-processing	-insufficient heat treatment to destroy microorganisms, residual carbohydrates in excess that permits secondary fermentation
	-excessive moisture and residual carbohydrate in non-cooked product. Insufficient drying rate
	-temperature abuse post-packaging
off-flavor	-microbial contaminant either growing during fermentation or post-packaging
	-use of spoiled meat raw materials
	-poor sanitation post-processing
	-chemical contaminant
slimy, gassy product in package	-yeast or heterolactic contamination post-processing
	-excessive moisture content
	-inadequate smoke concentration at product surface (i.e. semi-dry products)
green or gray coloration	-insufficient cure level or heat
	-oxidation of meat pigments via microbial contaminant, metal contaminants
	-exposure to sunlight
	-high pH
"greasing out"	-excessive heating rate
	-excessive fermentation temperature
	-unstable meat mix, low-binding meats
	-overworking raw meat mix

CHAPTER 5

Meat Product Safety and Shelf-Life

Introduction

Meat product safety and shelf-life are dependent on rapid preservation techniques instituted by man post-slaughter. In general, this preservation is accomplished by the synergistic effect of several methods rather than the use of a single procedure (Zottola, 1972). Historically, drying, salting, smoking, and fermentation have been used in combination to preserve meat. The addition of salt, sugar, nitrite, smoke, and the subsequent holding of the product at reduced temperatures with a decrease in the oxidation-reduction potential (i.e. casing), favors the growth of the lactic acid bacteria over the undesirable microorganisms. During growth, the lactic organisms ferment the sugar to primarily lactic acid, thus reducing the meat pH and providing prolonged stability against the proliferation of food pathogens. Lactic acid is non-toxic to man and moderately pleasant to the palate. Further, harmful bacteria tend to die off when stored in the finished product.

Safety

The prevalence of pathogens in raw meats and formulation ingredients has been studied extensively. Pathogenic bacteria, parasites, viruses, and/or mold spores may be present in sausage ingredients or they may come from the equipment, environment, or personnel in the plant.

In the United States, the prevalence of staphylococci in red meats and poultry at the retail level and in fresh pork sausage has been documented at over 50% (Genigeorgis, 1974; Surkiewicz et al., 1972). Fermented salami formulations (107 samples) in one area demonstrated a range of staphylococcal log counts from less than 2 to 4.75/g with a geometric mean of 2.6/g. Ninety-five percent of the samples had a log count of 2.4 to 2.7/gram (Genigeorgis, 1976). Some specific meat trimmings, such as pork cheeks, exhibit high numbers of staphylococci (Hill, 1972) and localized abscesses may contain millions of staphylococci that may survive processing and cause food poisoning

(NCDC, 1972). Generally, the contamination of spices and other non-meat ingredients is negligible due to previous processing (i.e. gas sterilization, pasteurization).

The occurrence of salmonellae in fresh meats varies considerably although only low numbers of certain strains are necessary to cause food poisoning. In one study, 560 fresh pork sausage samples yielded 28% that were contaminated with low numbers of Salmonella (Surkiewicz, 1975).

Due to its ubiquitous distribution in the environment, Clostridium perfringens would be expected in most meats, ingredients and the processing environment, although probably in low numbers.

Trichinella spiralis occurred between 0.125 and 0.51% in farm-raised and garbage-fed hogs, respectively, during 1966 to 1970 (Zimmerman and Zinter, 1971). During 1965 to 1969, 0.16% of bulk sausage samples and 0.5% of fresh link sausage examined were found to contain trichinae (Zimmerman, 1970).

The occurrence of Clostridium botulinum spores in fresh and semi-preserved meats is low (<1%) and when spores are detected, they are present in low numbers (<4/kg) (Riemann et al., 1972; Roberts and Smart, 1976).

Although extremely variable, most common food-borne pathogenic microorganisms can be encountered periodically and should be assumed present. The ability of food pathogens to survive, initiate growth, and produce toxins (in some cases) depends on their ability to overcome the inhibitory environment created during formulation and processing. Important components of the environment are 1) initial meat formulation and eventual changes in pH, brine, a_w, redox potential, and nitrite; 2) temperature, relative humidity, casing type, and rate of chemical and physical changes; 3) the initial numbers and types of pathogens; 4) and the numbers of competing microflora in the sausage formulation, including any starter cultures (Table 28).

Many fermented meat products are often held at elevated temperatures during processing to insure rapid fermentation, but these temperatures can also accentuate the growth of the pathogenic bacteria. In addition, many of these products are not fully-cooked, and they are usually eaten without further cooking by the consumer. These considerations make strict control of the product essential. Although proper sanitation, employee hygiene, and the control of raw materials definitely reduce contamination, the ultimate control of product safety must be inherent in the formulation and process. The use of starter cultures provides sufficient microbial numbers to insure numerical dominance over the natural flora, including pathogens, and in combination with the proper processing controls, guarantees the safety and quality of the final product.

Table 28. General growth characteristics of food poisoning bacteria and molds relevant to the processing of fermented meats (from Genigeorgis, 1976).

Bacteria or molds	Growth temperature range (C)	Lowest pH permitting growth	Maximum brine** concentration permitting growth (%)	Minimum water activity permitting growth
Salmonella spp	5.2-45	4.05	8	0.94
Staphylococcus aureus	6.7-45.6	4.0 ($+O_2$)	16-18 ($+O_2$)	0.83-0.86 ($+O_2$)
		4.6 ($-O_2$)	14-16 ($-O_2$)	0.9 ($-O_2$)
S. aureus enterotoxin				
production	10-45	4.0 ($+O_2$)	10 ($+O_2$)	0.90 ($+O_2$)
		5.3 ($-O_2$)	9.5 ($-O_2$)	0.94 ($-O_2$)
Clostridium perfringens	6.5-50	5.0	6	0.93-0.97
Bacillus cereus	7-49	4.4	7.5	0.955
Clostridium botulinum types				
A	10-48	4.7	10	0.93-0.95
B	10-48	4.7	10	0.93-0.94
B (nonproteolytic)	3.3	NK	NK	NK NK
E	3.3-45	5.0-5.4	5.3-5.5	0.94-0.97
F	3.3	NK	NK	NK NK
Molds*	-12-55	1.7	20	0.62
Mycotoxin production	4-40	1.7	10	0.8-0.85

NK = not known

*Extreme conditions at which at least one species was able to grow and produce a mycotoxin

$$\text{**Brine \%} = \frac{\text{NaCl (\%)}}{\text{moisture (\%)} + \text{NaCl (\%)}} \times 100$$

$$\text{Water activity} = \frac{\text{Vapor pressure of food}}{\text{Vapor pressure of distilled water of same temperature}}$$

Staphylococcus aureus. Staphylococcal food poisoning caused by defective, fermented dry or semi-dry sausage is a potentially significant problem according to a USDA task force report (USDA, 1977). Since 1967, there have been at least six publicized incidents of staphylococcal food poisoning traced to fermented sausage (USDA, 1977; NCDC, 1979). These incidents involved different major meat companies, and large amounts of product. In each case, there was a lack of scientifically-based controls designed to insure that the final product was safe for consumption. Further, the task force concluded that the current trends in production practices with increased volumes per facility and reduced processing times at higher temperatures may enhance the staphylococcal problem (Table 29). Specific recommendations, in addition to a continued emphasis on good raw materials and proper sanitation, include controlled acidulation of all fermented sausage with either microbial starter cultures or chemical acidulants (Figure 32). The National Academy of Sciences has also recommended that manufacturers add lactic starter cultures and/or chemical acidulants to inhibit staphylococci multiplication (NAS, 1975).

Staphylococcal food poisoning is caused by the ingestion of a heat-stable enterotoxin produced as a by-product during the growth of certain strains of Staphylococcus aureus (Riemann, 1969). Most food poisonings have involved Type A enterotoxin, which is also true in cases where fermented sausages have been implicated (Barber and Deibel, 1972). Although other meat and food products have been implicated in this type of food poisoning (Minor and Marth, 1972), some of the conditions involved in the manufacture of fermented sausage allow for the successful occurrence, competition, and growth of S. aureus strains. These formulation and processing parameters include the use of a high percentage of pork raw materials, a high salt content, reduced water activity, lack of natural smoke application to the surface, and higher incubation temperatures (Peterson et al., 1964; Surkiewicz et al., 1972; Tatini et al., 1976). In general, staphylococci proliferate and produce enterotoxin during the initial stages of the sausage fermentation and in the outer one-eighth inch of the product (Barber and Deibel, 1972). Final product pH, heat treatment, water activity, and/or refrigeration will not preclude staphylococcal food poisoning if the toxin was already produced in the initial fermentation stage. Proper control mechanisms in the successful production of fermented sausage must be applied to the initial formulation and processing stages (USDA, 1977).

The Good Manufacturing Practices (GMP's) for fermented dry and semi-dry sausage, recently developed by the Industry in the United States, particularly address those critical control points so as to minimize the opportunity of S. aureus reaching levels of public health significance. They state that "once the sausage pH reaches 5.3 or less, the environment for S. aureus is effectively controlled, thus minimizing the potential for growth to a dangerous level. During fermentation of sausages to a pH of 5.3, it is necessary to limit the period for which the sausage meat is exposed to temperatures exceeding 60F (15.6C) or higher". In addition to encouraging

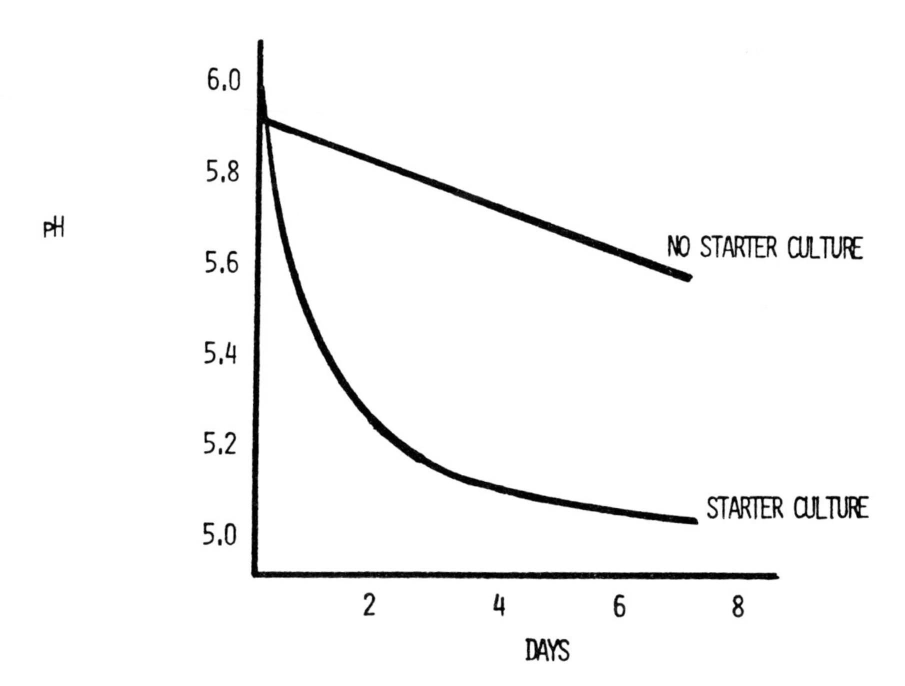

Figure 32. Fermented sausage - pH reduction at 85F (29.4C).

Table 29. Factors influencing the growth of S. aureus in fermented sausage (from USDA, 1977).

1. Increased sausage sales of seasonal or sales growth origin may dictate acceleration of a process and/or overcrowding of existing facilities.

2. Processing changes are often made gradually or abruptly without knowledge of their effects on the growth of S. aureus.

3. To produce fermented sausage, lightly salted meat is intentionally incubated to promote microbial growth. Unless controlled, S. aureus can grow to levels of tens of millions of cells on the surfaces of the sausage. Little or no growth occurs in sausages at depths 1/8-inch or below the surface.

4. S. aureus is a common bacterium in both man and animals. It is always in the chopped meat mix. Nearly half of all strains of S. aureus isolated from both human and animal sources have been found able to produce enterotoxin.

5. High localized levels of S. aureus in the meat trimmings do not appear to be a significant factor, possibly because of voluminous dilution in the sausage mix.

6. S. aureus does not grow well anaerobically in the presence of nitrite. This is one reason why S. aureus growth occurs as a ring phenomenon on sausages.

7. Staphylococci do not compete well with other bacteria. In raw meat held at 70F (21.1C), little or no growth occurs.

8. S. aureus is a true mesophile and grows fastest at 102F (38.9C). Naturally occurring lactic acid bacteria in meat grow best at 70-80F (21.1-26.7C). At temperatures above 75-80F (23.9-26.7C) their rate of growth is progressively decreased.

9. At temperatures above 80F (26.7C) "high temperature" strains of lactic acid bacteria are necessary to repress the growth of S. aureus. These must be present in large numbers to provide the necessary competition. Since such strains are not common to raw meat, they must be added in large quantities as pure cultures or by backslopping active-fermenting product. If these procedures fail for any reason, S. aureus can grow to levels capable of producing physiologically significant concentrations of enterotoxin.

10. During the fermenting of sausages, S. aureus growth and enterotoxin formation occurs during the first 3-5 days of processing, if at all. During drying, viable levels of S. aureus gradually decrease.

11. Initiation of death of S. aureus correlates with the low pH levels attained for the fermentation and the setting (firming) of the emulsion.

Table 29 (Continued)

12. Enterotoxin is heat stable, e.g. survives--minimum at 100C (212F) without inactivation.
13. At the end of the drying period, S. aureus levels may have fallen to 1/100,000 their level in the green room. Enterotoxin levels do not change apparently.
14. Sugar, particularly dextrose, as added to meat mixes, is the source of most of the lactic acid produced during fermentation. The final pH is in part related to the amount of sugar incorporated into the mix.
15. Chemical acidulation primarily through the addition of d-gluconolactone into the formulation is a commercial alternative to microbial fermentation. The compound dissociates to gluconic acid when warmed and hydrated.
16. Excessive amounts of salt, nitrite or other ingredients in the raw formulation may be detrimental to the lactic acid bacteria but not to the staphylococci. Appropriate formulations and thorough mixing of additives into the chopped meat, thus are important.
17. Sausage mixes in which lactic acid bacteria have been allowed to initiate growth at low temperatures appear to be highly resistant to the subsequent growth of S. aureus.
18. Potentially, "rework" or back inocula containing high levels of S. aureus could seed new batches of meat with active staphylococci and perpetuate or seed subsequent batches.

---

manufacturers to use controlled acidification procedures (mainly starter cultures), the GMP's also outline time-temperature-pH limits "which the industry has found to be safe, acceptable and attainable under practical production conditions" (Table 30).

Total numbers of coagulase positive staphylococci in the final product are not indicative of the presence of enterotoxin and the relative safety of the product (Niskanen and Nurmi, 1976). Since 40-50% of the staphylococcal strains in the environment produce enterotoxins, a potential problem also exists where large numbers of coagulase negative strains are detected (Johnston, 1980). A more rapid and accurate assay has been proposed as more indicative of the presence of enterotoxins. This method depends upon the detection of a heat-stable deoxyribonuclease which is associated with enterotoxin production (Emswiler-Rose et al., 1980).

Inhibition of staphylococcal growth is enhanced as the ratio of bacteria found in meats to staphylococci increases and the temperature of storage decreases (Peterson et al., 1962). Suppression of staphylococcal growth generally suppresses enterotoxin production

Table 30. Time-temperature control, dry sausage (adapted from AMI, 1982).

Constant temperature processes

Maximum degree/hours[1] to reach pH 5.3	Temperatures	Allowed hours by Guideline
1200	75F (23.9C)	80
1200	80F (26.7C)	60
1200	85F (29.4C)	48
1000	90F (32.2C)	33
1000	95F (35.0C)	28
1000	100F (37.8C)	25
900	105F (40.4C)	20
900	110F (43.0C)	18

[1]Degrees are measured as excess over 60F (15.6C). Degree/hours are the product of time in hours at a particular temperature and the "degrees".

(Niskanen and Nurmi, 1976) although the reverse is not true (Haines and Harmon, 1973). The beneficial effect of microbial starter cultures in inhibiting staphylococcal growth and enterotoxin production in fermented sausages has been readily demonstrated (Table 31). The large numbers of lactic acid bacteria provide a consistent controlled fermentation, accelerate the rate of acid development, and thus, indirectly retard the growth of staphylococci (Nurmi, 1966; Genigeorgis et al., 1971; Barber and Deibel, 1972; Daly et al., 1973). Additionally, certain microbial cultures also exhibit antagonistic effects on either staphylococcal growth and/or enterotoxin formation through the production of inhibitory substances (e.g. hydrogen peroxide, antibiotics) and/or a competition for essential nutrients (Oxford, 1944; Hirsh and Wheater, 1951; Troller and Frazier, 1963; Haines and Harmon, 1973).

Salmonella. The prevalence of salmonellae in meat varies considerably, although red meat and poultry do provide main vectors for its distribution (Goepfert, 1980). Recent incidences of food poisoning attributed to salmonellae in meat have involved raw hamburger (NCDC, 1975) and roast beef (Mayer et al., 1975; Checko et al., 1976; Jordan et al., 1976; Checko et al., 1977), where relatively low infectious doses have been demonstrated. Rarely have episodes of salmonellosis been attributed to fermented meats (Smith et al., 1975) although recent studies have established the ability of salmonellae

Table 31. Staphylococcal enterotoxin development, dry sausage, 22-24C (adapted from Niskanen and Nurmi, 1976).

Starter Culture	3 Days			7 Days		
	Log CPS	pH	Enterotoxin	Log CPS	pH	Enterotoxin
-	8.84	5.9	+	8.88	5.7	+
+	6.78	5.6	-	7.53	5.3	-

to survive some commercial processing conditions (Smith et al., 1975; Masters, 1979).

Lactic acid bacteria are definitely inhibitory to the growth of salmonellae with the relative effect dependent on the species, strain, ratio of lactics to salmonellae, the incubation temperature, and the degree and rate of acid production (Park and Marth, 1972). In semi-dry sausage fermented at 95F, the processing schedule demonstrating the slowest fermentation rate was the least restrictive to salmonellae survival. A corresponding process utilizing a commercial lactic acid starter culture reduced fermentation time and was the most restrictive to salmonellae survival while a natural lactic flora fermentation was the next most restrictive (Masters, 1979). Other studies have also demonstrated that a heavy lactic bacteria inoculum, achieved through the use of a starter culture, and high temperatures will suppress growth more rapidly than a light inoculum and low temperature (Peterson et al., 1964).

European workers demonstrated that starter cultures alone prevented the growth of S. seftenberg in dry sausage and, with higher initial contamination levels, starter cultures accelerated the decline in salmonella content (Sirvio and Nurmi, 1977). An independent investigation concluded that salmonellae survival in fermented sausages depends on the amount of contamination and whether starter cultures are used (Genigeorgis, 1976).

In addition to the rapid production of acid and lowering the product pH, the inhibition of salmonellae growth and survival by lactic acid bacteria can also be due to the production of other compounds (Goepfert and Chung, 1970; Genigeorgis, 1976). This is especially true in nonfermented meat products (Reddy et al., 1970; Daly et al., 1972).

Clostridium botulinum. Botulinal toxin formation in meats is influenced by oxidation-reduction potential, pH, water activity, salt, nitrite, moisture level, and temperature (Baird-Parker and Freame, 1967; Johnston, 1969; Christiansen et al., 1973; Roberts and Ingram, 1973; Bowen and Deibel, 1974; Collins-Thompson et al., 1974). During the past decade, the nitrite controversy in cured meats has been the basis for further research into the factors affecting the growth and toxin production of Clostridium botulinum. The role of the lactic acid bacteria in inhibiting the growth of C. botulinum in meat has been readily demonstrated (Table 32). Lactic starter cultures, in combination with either sucrose or dextrose, have proven effective through rapid acid production in preventing toxin production even in the absence of nitrite (Christiansen et al., 1975; Tanaka et al., 1980). Some investigators have postulated that nitrite levels may be safely lowered in certain meat products if lactic cultures and carbohydrates are added (Tanaka et al., 1980). The inhibition due to the lactic acid bacteria and the presence of a fermentable carbohydrate and nitrite probably explains the past safety of fermented sausage with respect to botulism (Jay, 1970).

Trichinae spiralis. Federal regulations dictate the formulation and processing parameters of meat products to safeguard against T. spiralis (USDA, 1973). This includes freezing, formulations, heating and dehydration. Microbial starter cultures indirectly contribute to more rapid trichinae destruction in fermented, uncooked sausage through a rapid, consistent and controlled fermentation. This allows for a consistent dehydration rate and increases the toxicity of added nitrite.

In addition to providing an effective control mechanism for the more common meat pathogens, selected microbial cultures also have exhibited inhibitory action against other undesirable species including bacilli, gram negative enterics, and yeasts (Nurmi, 1966; Hurst, 1973; Brankova, 1976). Trials in fermented sausage demonstrate that added starter cultures effectively control enterococci and enterobacteria (Liebetrau and Grossman, 1976).

Table 32. Botulinal toxin development, summer style sausage @ 27C (adapted from Christiansen et al., 1975).

Nitrite (ppm)	Starter Culture	Dextrose	Toxic/25
0	-	+	8
0	+	-	22
0	+	+	2
50	+	+	0
150	-	-	14
150	+	+	0

Histamine Control. Histamine has been associated with several outbreaks of food poisoning in the United States and numerous episodes abroad (Taylor et al., 1978). Histamine is formed in foods by the bacterial decarboxylation of histidine and is a normal constituent of fermented foods such as cheese, wine, and sauerkraut (Taylor et al., 1978), and other food products exposed to microbial degradation. Histamine concentrations in fermented meats can be variable and dependent on the length of the aging process (Cantoni et al., 1974; Rice et al., 1975; Taylor et al., 1978). Higher histamine levels have been found in dry sausages where a natural fermentation process is employed for an extended aging period. Apparently, microbial contaminants in the naturally-fermented products account for the higher levels of histamine detected (Taylor et al., 1978). The addition of commercial lactic cultures appears to effectively prevent histamine accumulation through control of the natural fermentation. Similar results have demonstrated the use of a starter culture greatly decreases the possibility of microflora development that possess both tyrosine decarboxylase activity and the proteolytic activity required to produce potentially hazardous tyramine concentrations (Eitemiller et al., 1978). Several reports have confirmed the lack of histamine and tyramine production by commercially available *Lactobacillus* and *Pediococcus* meat starter cultures (Rice and Koehler, 1976).

Shelf-life

Fermented meat products have traditionally demonstrated an extended shelf-life through a combination of reduced moisture content and pH. The United States Department of Agriculture recognizes sausage having a moisture/protein ratio of 3.1 or less, and a pH of 5.0 or less, as not requiring refrigeration (USDA, 1977). Additionally, shelf-stable meat products are classified as having a pH at or below 5.2 and a water activity at or below 0.95, or a pH below 5.0 or a water activity below 0.91 (Leistner and Rodel, 1975). The shelf-life of these products is generally not limited by bacterial deterioration, but by chemical or physical spoilage.

Microbial cultures contribute to the shelf-life of fermented meats mainly by consistent and controlled acidification. The acid development inhibits undesirable microorganisms and allows efficient dehydration. Specific micrococci cultures also enhance cured color stability and prevent rancidity development through the reduction of peroxide formation via a catalase system (Andres, 1977). Certain yeast cultures of the Debaryomyces family have also been shown to accelerate and stabilize the color development at the surface of dry sausages (Coretti, 1977).

Fermented meat technology with the addition of starter cultures is also being applied to fish products to extend shelf-life (Herborg and Johansen, 1977; Schubring and Kuhlman, 1978).

Non-fermented meats

Microbial cultures also have been added to non-fermented meat-type products to prolong shelf-life and inhibit pathogenic microorganisms. Selected cultures have been applied successfully to boiled ham, ground beef, bacon, beef cuts, frankfurters, dry-cured hams, mechanically deboned poultry, smoked turkey, and shrimp (Pina, 1977; Reddy et al., 1970; Tanaka et al., 1980; Tezcan and Yuecel, 1975; Petaja, 1977; Bartholomew and Blumer, 1977; Raccach and Baker, 1978; Moon et al., 1980). In lieu of acid production, these preservation systems rely on inhibitory substances produced by the starter cultures, including hydrogen peroxide, antibiotics, and as yet unidentified growth inhibitors (Dubois et al., 1979). Hydrogen peroxide (Tables 33 and 34) appears to be the major inhibitor produced by the lactobacilli cultures that have been studied (Gilliland, 1980). Most research indicates that large numbers of lactobacilli are required to effect significant inhibition, and further studies are required, prior to commercialization, to understand the parameters in controlling the amount of inhibitory action.

Microbial additives to non-fermented products also can serve as a "safety factor" to prevent food pathogen growth and toxin formation if the products are temperature abused. When the product is held refrigerated, the added cultures have limited metabolic activity, but when temperature-abuse occurs (inadvertent warming), the microorganisms are activated with the fermentation of added sugars to acid preventing the growth of food pathogens (Tanaka et al., 1980).

Nutritional considerations

The United Nations World Hunger Program confirmed at a recent workshop that food fermentation can be more important as a means of food preservation and food conservation (UN, 1979). Through the use of specific microorganisms, perishable foods as rice, wheat, sorghum, corn, soybeans, black gum, cassava, taro, milk, eggs, fish, and meat can be converted to consumer products with better nutritive value, digestibility, appearance, and shelf-life than the original raw materials. Various agricultural waste products can also be converted into palatable foods through controlled fermentation. The workshop recommended future research include more work with the fermentation of indigenous foods to permit higher consumption.

Although the direct contribution of microbial cultures to the nutritive value of meat products has not been extensively studied, the prolonged stability of fermented meats certainly allows for greater consumption of a perishable raw material. This natural preservation system also precludes alternative means of preservation such as extreme heat and chemicals that may reduce nutritive value. The higher protein value of many fermented meats (Kiernat et al., 1964) generally results from the drying process that is consistently achieved through controlled fermentation.

Table 33. Peroxide production by Lactobacilli and Pediococci at 5C (41F)[a] (Gilliland and Speck, 1975)

Organism	H_2O_2 (OD at 400 nm)[b] 10% NFMS[c]	Broth[d]
L. bulgaricus NCS1	0.017	0.791
L. bulgaricus HWD	0.045	0.732
L. lactis BYL1	0.243	2.000
P. cerevisiae	0.007	0.007

[a]With continuous agitation: 22 hr. incubation.

[b]The higher the optical density, the more peroxide is present.

[c]Nonfat milk solids: autoclaved (15 min. at 121C, 250F).

[d]0.25% beef extract + 0.1% glucose: autoclaved (15 min. at 121C, 250F).

Table 34. Effect of cells of L. bulgaricus, L. lactis and P. cerevisiae on growth of psychrotrophic bacteria in ground beef[a] at 5C, 41F (Gilliland and Speck, 1975).

Days at 5C	Control	L. bulgaricus[b] NCSI	L. lactis[b] BYL1	P. cerevisiae[b]
0	6.7×10^{2}[c]	8.3×10^{2}	8.5×10^{2}	8.4×10^{2}
3	1.5×10^{5}	5.3×10^{3}	4.7×10^{3}	1.5×10^{4}

[a]Ground beef prepared in laboratory.

[b]Approximately 5×10^8 cells added per gram.

[c]Colony counts per g on CVT agar.

CHAPTER 6

Nitrosamine Reduction in Bacon

Nitrite

Nitrite (NO_2^-) is added to meat products to effect cured-product stability, flavor, and color (CAST, 1978). It was recognized first as a curing agent when nitrate (NO_3^-) was identified as an impurity in the salt used during the traditional process of "salting" meat to preserve it against microbial degradation (Haldane, 1901). The nitrate impurity was being microbially reduced to nitrite and nitric oxide (NO).

(1) $\text{Nitrate} \xrightarrow[\text{organisms}]{\text{nitrate reducing}} \text{Nitrite}$

(2) $\text{Nitrite} \xrightarrow[\text{absence of light and air}]{\text{favorable conditions}} \underset{\text{nitric oxide}}{NO} + \underset{\text{water}}{H_2O}$

(3) $\underset{\text{nitric oxide}}{NO} + \underset{\text{myglobin}}{Mb} \xrightarrow[\text{conditions}]{\text{favorable}} \underset{\text{metmyoglobin}}{\text{NOMMb--nitric oxide}}$

(4) $\text{NOMMb} \xrightarrow[\text{conditions}]{\text{favorable}} \underset{\text{nitric oxide myoglobin}}{\text{NOMb}}$

(5) NOMb + Heat + Smoke ⟶ NO-Hemochromogen
nitrose-hemochromogen
Stable pink pigment

Nitric oxide stabilizes meat pigment myoglobin through a reversible chemical bond in the same manner that muscle pigment is stabilized by molecular oxygen in the oxygenated postmortem meat system. Stabilized myoglobin causes a reflectance of light characteristic of cured meat color (Dryden and Birdsall, 1980). In 1925, the direct use of nitrite was approved for red meat products because of its importance as a coloring agent. Since its approval, nitrite has been shown to function in the curing process as a microbial preservative, flavoring agent, and antioxidant (CAST, 1978).

Clostridium botulinum is the most deadly pathogenic microorganism that could contaminate cured meats. The inhibitory effect of nitrite against C. botulinum is well documented (Christiansen et al., 1974; Roberts and Smart, 1974). Although the exact mechanism of the inhibition is not known, several researchers have postulated that nitrite inhibits C. botulinum toxin production by inhibiting spore outgrowth and/or by forming an inhibitory compound when nitrite is heated in meat (Sofos et al., 1979). Nitrite has also been shown to exert an inhibitory effect against C. perfringens (Buchanan and Solberg, 1972) and Staphylococcus aureus (Rhia and Solberg, 1975), two food pathogens often associated with processed meats.

The antioxidant and flavoring roles of nitrite appears to be interrelated (Freeman et al., 1982). Although evidence indicates that nitrite contributes to "cured meat flavor" (Mottram and Rhodes, 1973), the compound(s) responsible or the role of nitrite is unknown. Therefore, it has been postulated that the role of nitrite in regard to meat flavor is antioxidant in nature (Bailey and Swain, 1973). Apparently, the presence of nitrite inhibits the oxidation of lipids while in the absence of nitrite the formation of offensive carbonyl compounds by lipid oxidation results in a rancid or "warmed-over" flavor (Scalan, 1975).

Nitrosamines

Nitrosamines have been recognized for some time as a potential human health hazard due to their carcinogenicity in laboratory animals (Magee and Barnes, 1956). Nitrosamines are formed in certain foods, such as nitrite cured meat, when nitrite forms nitroso groups (-N=O) which can chemically bind to an amine nitrogen atom in certain organic compounds such as secondary amines. Dimethylnitrosamine and nitrosopyrrolidine are found most consistently in some nitrite-cured bacon after frying. An average of 9.3 ppb and a range of 1.7 to 21.8 ppb of nitrosopyrrolidine have been reported in 12 samples of fried bacon (Sen et al., 1979). Earlier reports indicated that the concentration of nitrosopyrrolidine in fried bacon ranged from 4 to 108 ppb (Fazio et al., 1973; Sen et al., 1973). The occurrence of nitrosamines in other cured meat products has been sporadic although dried-cured hams and bacon may also pose a problem (USDA, 1980).

Continuous consumer concern over the carcinogenicity of nitrosamines found in fried bacon almost led to a governmental ban in the United States in 1980 on the use of nitrites in meat processing. Since approximately 60% of the pork produced in the United States is processed with nitrite (CAST, 1978), a complete ban would be an economic disaster to the pork industry. In addition, "cured meats" would be practically eliminated, since a total replacement for the functions of nitrite is unknown. A nitrite ban also could increase conceivably the danger of food poisoning, such as botulism, and definitely reduce the shelf-life that is characteristic of many processed, cured meats such as frankfurters, bologna, salamis, ham, etc.

In an attempt to reduce the levels of nitrosamines found in fried bacon, the USDA required in June, 1978, that bacon be produced using a "reduced" nitrite level of 120 ppm and 550 ppm sodium ascorbate or sodium erythorbate (Fed. Reg. 1978). Various studies had shown that bacon produced with these initial nitrite and ascorbate levels (i.e. reducing agent) had negligible levels of nitrosamines (<10 ppb) and are safe from botulinal hazards (Greenberg, 1975). However, under commercial conditions, these levels of nitrite and ascorbate proved only partially effective, and nitrosamine levels greater than 10 ppb are still being found routinely in fried bacon.

Nitrosamines are formed primarily by the action of nitrous anhydride (N_2O_3) on an unpronated amine with the subsequent release of nitrite ion (Tannenbaum and Fan, 1973). Since nitrous anhydride is formed from two nitrous acid molecules, the rate of nitrosation is second order with respect to the nitrous acid concentration and first order with respect to amine concentration. This reaction is highly pH dependent, because a pH drop causes the nonionized HNO_2 to increase and the nonionized amine to decrease. The optimum pH for nitrosation is approximately 3.5 (pKa HNO_2=3.36) since below this pH the principle effect is decreasing the nonionized amine concentration (Mirvish, 1970). As a result, the ease of nitrosation of amines has an inverse relationship with amine basicity. Other factors relating to meat systems which affect the rate of nitrosation include temperature and sodium chloride. In bacon without added phosphate, the pH is approximately 5.6, and high temperatures (i.e. frying) are generally required to effect nitrosamine formation (Pensabene et al., 1974). Cooking temperatures below 100C failed to produce nitrosopyrrolidine, but increasing temperatures resulted in higher concentrations of nitrosopyrrolidine, reaching a maximum level at 204C. In bacon with added phosphate (as is typical of commercial practice), the pH range is generally 6.2-6.5, increasing the potential for nitrosamine formation since the residual nitrite concentration is more stable at this pH. Higher levels of sodium chloride (i.e. 3.5%) have been shown to reduce residual nitrite, thus lowering the potential for nitrosamine formation (Lee and Cassens, 1980). These researchers concluded that the higher ionic strength, due to the sodium chloride, causes a decrease in pH. In addition, the increased availability of the proteins in solutions high in sodium chloride make them more susceptible to react with nitrite and thereby lower the residual concentration.

Nitrosamine formation in cured meats can be prevented if nitrite is not utilized in the process (Sen et al., 1974). To reduce nitrosamine concentrations, sodium ascorbate or its isomer sodium erythorbate is widely used in the meat industry (Fiddler, 1973). Although the mechanism is not fully documented, the ascorbate either reduces nitrite, via a nitrosated intermediate, to nitrous acid which does not react with the amines, or it competes with the amines for the available nitrous anhydride. At pH 5.6, the probable pH of most cured meats (without added phosphate), ascorbic acid is more readily nitrosated than amines. Numerous other compounds, that are capable of reacting with nitrite, have been shown to partially reduce nitrosamine formation (Fox and Nicholas, 1974; Gray and Dugan, 1975a). The antioxidants, α-tocopherol, propyl gallate, and butylated hydroxytoluene all have shown inhibition in model systems; however, their water insolubility limits their effectiveness in actual bacon manufacturing.

Reducing the residual nitrite content in bacon will minimize subsequent nitrosamine formation, but it will also increase the potential for botulinal toxin formation. A system was required that would reduce nitrite without sacrificing safety.

Lactobacilli Cultures

Microbial starter cultures composed of lactobacilli were introduced to bacon processing to reduce the residual nitrite content while maintaining the safety of the product (Bacus, 1978). In combination with a fermentable carbohydrate, the lactic acid culture reduces the bacon pH during typical smokehouse processing and/or during subsequent refrigerated storage. The lower pH dissipates residual nitrite content, and both factors can reduce potential nitrosamine formation at the time of frying. Bacon pH and the residual nitrite content are two of the primary factors in determining potential nitrosamine formation during a typical frying process.

As a result of the test data submitted by ABC Research and since the safety and suitability of lactic acid microorganisms had been established and their use approved under the Federal Meat Inspection Act in dry and semi-dry sausages (9CFR-318.7), the USDA amended the Federal meat inspection regulations in 1979 (Houston, 1979) by "allowing the use of acid producing microorganisms such as lactobacilli...in the processing of bacon for the purpose of lowering the pH in order to dissipate residual nitrite and reduce nitrosamine formation" (Table 35).

The occurrence of nitrosamines, primarily in fried bacon, can be attributed to the basic chemistry of the raw pork bellies, the inconsistency of the raw material and processing techniques, and the high frying temperatures encountered during the traditional preparation of the product for consumption (Tables 36-38). Representative sampling and monitoring/control of nitrosamine levels has proven very difficult since bacon is not a formulated product, rather slices are obtained from a single portion from a single belly from a single animal. In commercial production, the processing equipment utilized to inject the bellies with the curing brine is not extremely accurate, and each belly may possess different characteristics that affect absorption of the brine and retention during smokehouse processing.

In the "cultured" process, a strain of _Lactobacillus plantarum_ is added directly to the curing brine, or "pickle", as to achieve 10^6 microorganisms per gram of sliced bacon. Depending upon the "pumping yield" and "smokehouse yield" of the respective product, the initial

Table 35. Approval of substances for use in preparation of product. Meat and poultry inspection regulations (Houston, 1979).

Class of substance	Substance	Purpose	Products	Amount
Flavoring agents: protectors and developers.	Harmless bacteria starters of the acidophilus type, lactic acid starter or culture of *Pediococcus cerevisiae*.	To dissipate nitrite.	Bacon	Sufficient for purpose.

Table 36. Variable factors affecting nitrosamine levels in fried bacon.

1. Hog - type, size
   - growth conditions
   - season slaughtered
2. Age of bellies - frozen vs. fresh
   - time interval
3. Belly - size
   - lean vs. fat
   - uniformity
4. Curing brine - composition
   - age
   - pH
   - temperature
5. Processing - yield, pumping percent
   - distribution
   - drain
   - temperatures, times
   - smoke
   - chilling
   - packaging, storage, time
   - slice thickness
6. Analytical sample - representative
   - composition
   - age
   - temperature
7. Analytical procedures - frying temperatures
   - sampling
   - background

Table 37. Bacon pump yields - paired bellies.[1]

Raw weight (lb)	Number of bellies	Low	High	Average
10-12	24	5	15	11
10-12	24	7	18	13
14-16	22	12	20	16
14-16	22	9	20	15
16-18	20	12	18	14
16-18	20	10	17	14

[1] Commercial production test. Yield = weight belly after pump - raw weight ÷ raw weight x 100.

Table 38. Bacon belly-typical analyses after processing.

Analysis	Shoulder end	Central	Flank end
Salt %	1.9	1.6	1.8
Protein %	9.8	6.3	7.2
Fat %	55.5	64.4	63.2
Nitrite (ppm)	46	29	39

level of microorganisms added to the curing brine is between 10^8 - 10^9 per ml brine. Although the relatively high concentrations of salt and nitrite in the brine are toxic to microorganisms, viability studies demonstrated good stability over the time interval (8 hrs) that could be expected in commercial practice (Table 39). It was noted also that nitrite concentration and brine pH remained fairly constant in culture-added brine. These findings are significant since microbial survival in the brine is required to effect the final result and, on the contrary, if the microorganisms survive and begin producing lactic acid, the brine might become too acid as to allow premature nitrite loss. An important consideration in bacon processing as to reduce nitrosamine formation is that the initial nitrite be allowed to effect the desired curing reaction, and the residual nitrite be dissipated prior to frying. Premature dissipation of nitrite will preclude the desired flavor and color development.

Table 39. Curing brine stability (adapted from Ford, 1980).

Time (hr.)	APC (log 10/ml)[2]	pH	NO_2^- (ppm)
0	8.0	8.5	963
4	8.1	8.47	975
8	7.9	8.44	975
12	7.6	8.34	963
16	7.0	8.07	970
24	6.6	7.89	975

[1]Formulated brine ingredients: 11.54% salt, 3.85% sucrose, 0.42% sodium erythorbate, 0.09% sodium nitrite. Held at ambient temperature.

[2]APC = aerobic plate count.

The presence of a fermentable carbohydrate is also necessary to effect an adequate pH decrease in the bacon. Although most processors typically utilize either sucrose or dextrose in curing brines, a final level of 0.3-0.5% in the bacon is suggested. Sucrose is preferred to dextrose to minimize "charring" during frying (Bacus, 1978).

Sodium phosphates are used commonly in curing pickles to buffer the water pH and to provide higher processing yields, afford better slicing characteristics and reduce "splattering" during the frying of bacon. The use of phosphate will increase the pH in most meat products and provide a "buffer" to retard any pH decrease via starter cultures. Depending upon the respective levels of phosphate and a fermentable carbohydrate, the culture fermentation will "overcome" the buffer, but the pH decline will proceed more slowly. Generally, a cultured process will eliminate the phosphate or reduce the level, to expedite the fermentation.

After the raw pork bellies have been pumped with the curing brine, they are heat processed. The processing procedures (i.e. time, temperatures, smoke application) can vary between plants, although most manufacturers utilize a smokehouse where the bellies are heated from 4 to 22 hours at temperatures ranging from 110F to 140F (43.3C to 60C). During the heating cycle, natural or liquid smoke may be applied depending upon the color and flavor desired. To achieve the optimum slicing characteristics, most bellies are heated only to 124-128F (51 to 53C) internal temperature. Lower finish temperatures generally do not coagulate sufficiently the meat tissue (i.e. firmness) while higher temperatures overcook the tissue and can result in "shattering" the belly on a high-speed slicer (i.e. low slicing yields).

The relatively low, final temperatures attained in the belly allow the starter culture to survive the process. Depending upon the smokehouse temperatures and time cycles, the culture may, or may not, exert significant metabolic activity during the heat processing phase. Most lactobacilli cultures will not be active at product temperatures in excess of 105F (40.6C). However, some processors hold the pumped bellies at lower temperatures prior to smoking and the "come-up-time" may vary depending upon the "load" in the house, the air circulation, the size of the belly, and the initial temperature. Holding the bellies at lower temperatures prior to smoking can affect both the pH and the residual nitrite content of the resulting bacon (Table 40).

Generally, significant metabolic activity (i.e. pH reduction) is only evidenced during smokehouse processing if the holding time prior to smoking is extensive and/or the heating cycle favors the activity of the culture. Typical commercial processes (6-12 hrs. @ 120-140F, 47-60C) do not result in a significant pH reduction during the heating phase, but allow the culture to survive the process and function during subsequent distribution and storage.

A comparison of two heating cycles for cultured bacon (i.e. "matched" bellies) demonstrates that a shorter cycle can achieve ultimately a lower pH after refrigerated storage, presumably due to a higher survival rate and/or less injury to the culture (Table 41).

Table 40. The effect of holding "cultured bellies" prior to smoking (adapted from Ford, 1980).

	Control[1]	Held 20 hr @ 75F (23.9C)
APC ($\log_{10}$/gram)		
at slicing	7.12	7.93
at 21 day storage	8.20	7.88
pH		
raw	5.67	5.71
before smoke	5.75	5.65
at slicing	5.93	5.80
at 21 day storage	5.54	5.30
residual nitrite (ppm)		
before smoke	38.1	26.3
at slicing	31.7	11.5
at 21 day storage	7.0	6.8

[1]Smokehouse processed immediately after pump.

Table 41. The effect of the heating cycle on the characteristics of cultured bacon (adapted from Ford, 1980).

	Smokehouse cycle[2]	
	4.5 hr	12 hr
APC (log_{10}/gram)		
at slicing	7.36	7.34
at 21 day storage	8.19	7.75
pH		
raw	5.98	5.95
after pump	6.07	6.06
at slicing	5.93	5.86
at 21 day storage	5.40	5.62
residual nitrite (ppm)		
after pump	56.0	61.1
at slicing	73.6	33.9
at 21 day storage	6.5	8.3

[1]Both cooked to 124-126F (51-52C) internal temperature.

[2]4.5 hr. cycle = 3.5 hr @ 130F (54.5C), 1.0 hr @ 130F (54.4C) 100% RH

12 hr cycle = 3 hr @ 110F (43.3C)
3 hr @ 120F (48.9C)
3 hr @ 125F (51.7C)
3 hr @ 135F (57.2C)

This exemplifies the point that the greatest effect from the culture in a rapid process occurs during retail storage.

Several reports have indicated that higher levels of nitrosopyrrolidine (NPYR) are found in the adipose tissue, than in lean portions of bacon when fried separately (Fiddler et al., 1974; Mottram et al., 1977). This also correlates with research that has found bacon with a higher fat proportion to produce more NPYR on frying than leaner cuts such as ham, loins, etc. (Fazio et al., 1973). A comparison of lean and adipose tissue from "matched" bellies processed with or without starter culture demonstrates lower pH and residual nitrite content in both tissues after refrigerated storage of the sliced bacon (Table 42). The pH of both tissues from uncultured bacon remained relatively constant with time while the cultured tissues declined in pH. Since both bellies had similar initial values at slicing, the data again demonstrates the effect of the starter culture during

Table 42. Comparison of pH and nitrite concentration (ppm) of lean and adipose tissue from "paired" bellies during a bacon process[1] (adapted from Ford, 1980).

	Control (NO_2^-)	Cultured (NO_2^-)
Adipose tissue		
raw	6.37	6.37
after pump	6.27 (95)	6.43 (90)
at slicing	6.28 (48)	6.21 (47)
at 21 days	6.33 (20)	5.79 (8)
Lean tissue		
raw	6.04	5.99
after pump	5.98 (94)	5.99 (78)
at slicing	6.09 (97)	5.97 (101)
at 21 days	6.06 (37)	5.72 (8)

[1]Pumped 113%, 4 hr smokehouse process.

subsequent storage. A comparison of nitrosamine content demonstrated lower mean values in the cultured adipose tissue while no difference was observed in the lean tissue, presumably due to the low initial values (Table 43).

Commercial Use of Starter Culture

In the commercial application of starter cultures to bacon processing, no major changes are required by the manufacturer. The culture is added directly to the curing brine and the bellies are processed normally. Recommendations to enhance the effect of the culture are minimizing any phosphate usage and achieving 0.3-0.5% sucrose or dextrose in the sliced product. Extended smoking/heating schedules (i.e. in excess of 12 hrs.) are not recommended, although most processors are well within this time frame. To achieve the maximum effect, a residual microbial population of 10^5-10^6 per gram is desirable, thus final product temperature should not exceed 128-130F, 53.3-54.4C (as is generally the case). Composite data from various production plants has demonstrated the effectiveness of the lactobacilli in reducing nitrosamine formation while not adversely affecting the quality of the final product (Table 44).

Table 43. Comparison of nitrosamine content (ppb) of bacon tissues (from Ford, 1980).

	at slicing		at 21 days	
	Control	Culture	Control	Culture
Adipose tissue				
NPYR[1]	44.1	25.1	42.3	22.9
DMNA[2]	6.7	4.3	5.2	2.6
Lean tissue				
NPYR	4.6	4.7	-	-
DMNA	1.2	1.7	-	-
Drippings				
NPYR	19.5	14.5	10.5	6.1
DMNA	4.8	5.3	2.0	1.4

[1]Nitrosopyrrolidine

[2]Dimethylnitrosamine

Table 44. A composite of bacon analyses from commercial plants after 21 days refrigerated storage.[1]

	Residual nitrite (ppm)	SPC[2]	pH	nitrosopyrrolidine
Control	20-40	10^4-10^5	6.0-6.4	10-30
Cultured	4-16	10^6-10^7	5.3-5.6	2-9

[1]Various formulations and processing parameters

[2]Standard plate count

Botulinal Control

It has been well documented that sodium nitrite affords anti-botulinal protection to cured meat. A process that reduces the nitrite content of bacon will minimize nitrosamine formation, but it will increase the potential for botulinal toxin formation if the bacon is subjected to temperature abuse (Tanaka et al., 1980).

Botulinal toxin formation is influenced by pH, water activity, salt, nitrite, inoculum level and temperature (Baird-Parker and Freame, 1967; Collins-Thompson, 1974; Roberts and Ingram, 1973).

Different food systems depend on one or more of these parameters to afford protection, and these factors cannot be readily manipulated without affecting the safety and organoleptic quality. Inoculated pack experiments have demonstrated that toxicity development in bacon is erratic and nontoxic samples have had reduced pH values (Tanaka et al., 1980). The naturally-occurring, lactic microflora of bacon is also erratic and any reduced-pH products are due generally to their fermentative action. The control of the lactic microflora in meats via starter cultures has been effective for pH control and the resulting stability and safety of the products. Lactic bacteria and sugar have prevented botulinal toxin formation in fermented sausage (Christiansen et al., 1973).

The use of lactobacilli in combination with sucrose to control residual pH and prevent botulinal toxin formation in bacon has been very effective (Tanaka et al., 1980). Various experiments have demonstrated that lactobacilli + sucrose are an effective inhibitor of toxin formation in temperature-abused bacon with or without sodium nitrite (Figure 33) (Table 45). This inhibition is due to pH reduction, and both the culture and a fermentable carbohydrate (0.5%) must be present. Apparently, the use of sodium tripolyphosphate (0.31%) does not preclude the inhibition since no significant variation in pH was observed in one study (Figure 34).

Lactobacillus plantarum has been the starter culture employed in most bacon studies focusing on botulinal control and nitrosamine reduction. The salt resistance, fermentative ability, commercial availability and prior USDA-approval for use as a sausage culture were the primary factors associated with the selection of this microorganism. The selected strain also demonstrates good survival and little growth through refrigerated storage of the bacon. This provides a system in which properly handled bacon is not subject to extreme acid flavor, but the microorganism is available for rapid acid production should the product undergo temperature abuse. The culture will effect sufficient pH reduction in refrigerated bacon to reduce the potential for nitrosamine formation, but the ultimate pH attained, normally 5.3-5.6, is not low enough that detectable acid flavor results. The extent of acid produced can also be controlled through the amount of carbohydrate added.

Subsequent studies have examined various strains of pediococci, streptococci, and bacilli in an attempt to find more heat resistant microorganisms that can tolerate extended smoking/heating schedules for bacon (Tanaka, unpublished data). Although heat resistant strains are available, none of the microorganisms tested are capable of sufficient metabolic activity in a typical bacon system containing salt, nitrite, and high fat. *L. plantarum* strains still appear to be the most desirable starters for bacon although their use is somewhat limited to more conventional processes not in excess of 12-14 hours nor product temperatures in excess of 130F (54.4C).

Figure 33. Effect of sodium nitrite, sucrose and _L. plantarum_ on toxin formation by _C. botulinum_ and pH in bacon (from Tanaka et al., 1980).
Three samples were tested for toxicity at each sampling period. Open columns indicate number of toxic samples found at each sampling period. Closed circles indicate pH values determined.

Panel	Variables		
	$NaNO_2$ (ppm)	Sucrose (%)	_L. plantarum_ inoculated
A	0	0	+[a]
B	0	0.9	-[b]
C	0	0.9	+
D	120	0	-
E	120	0	+
F	120	0.9	-
G	120	0.9	+

[a]+, inoculated; [b]-, uninoculated

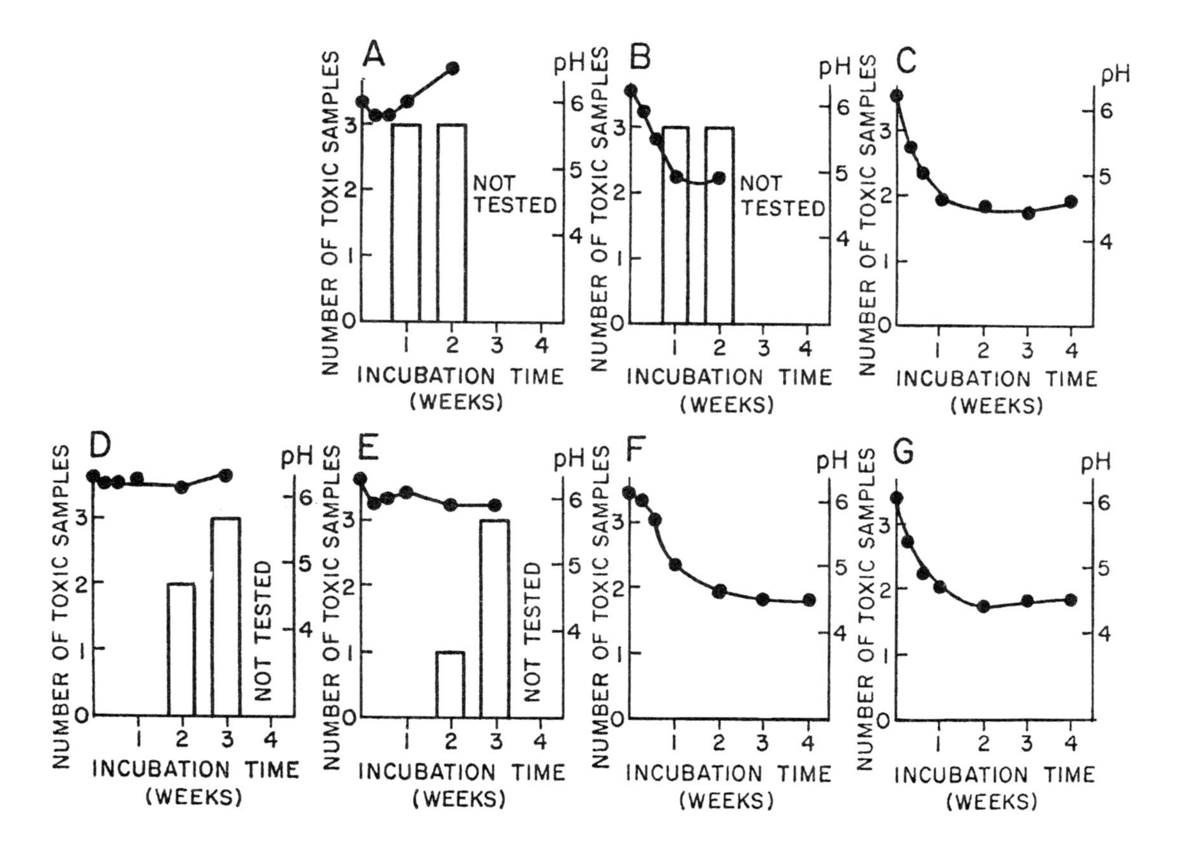
A
B
C
D
E
F
G
NUMBER OF TOXIC SAMPLES
INCUBATION TIME (WEEKS)
pH
NOT TESTED

Figure 34. Effect of variation in sucrose concentration and addition of STPP on toxin formation and pH in bacon inoculated with *L. plantarum* and *C. botulinum* (from Tanaka et al., 1980).

All the bacon was made with 120 ppm sodium nitrite.

Five samples were tested for toxicity for each lot at each sampling period. Open columns indicate number of toxic samples at each sampling period. Closed circles indicate pH values.

Panel	Variables	
	Sucrose (%)	STPP (%)
A	0.9	0.31
B	0.5	0.31
C	0.5	0
D	0.1	0

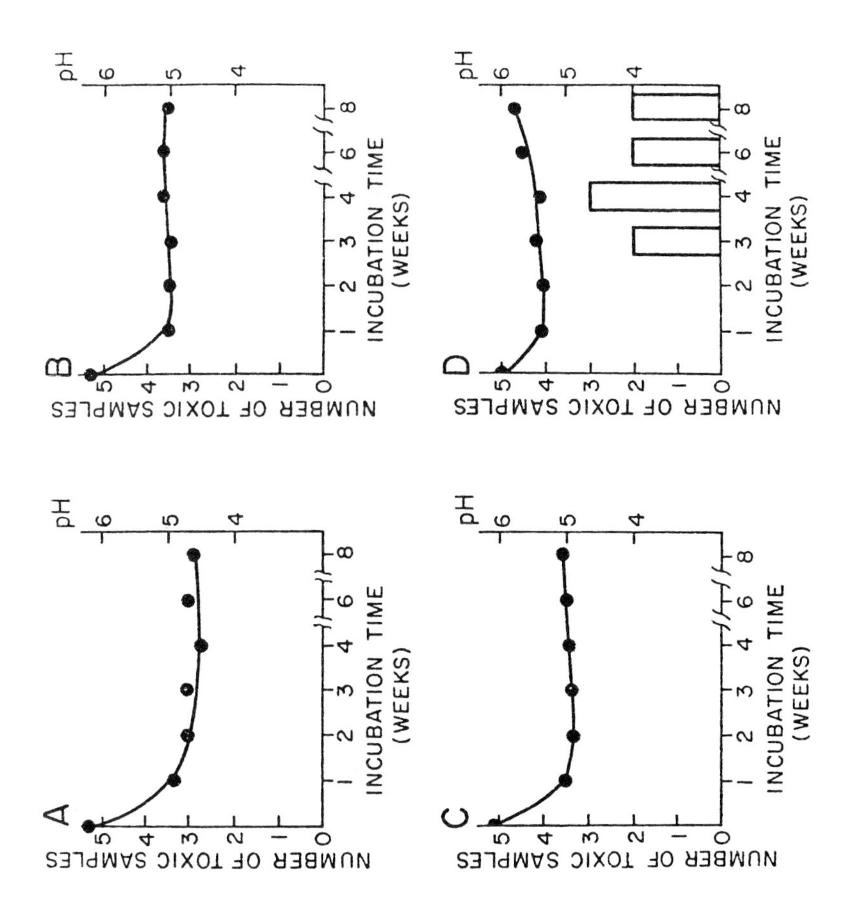
A
B
C
D
NUMBER OF TOXIC SAMPLES
INCUBATION TIME
(WEEKS)
pH

Table 45. Summary of toxicity results (from Tanaka et al., 1980).

Variables			Results of toxin assays	
$NaNO_2$ (ppm)	Sucrose[a] added	Lactobacilli added	Total number of samples tested	Number of toxic samples
0	-	+	27	26
0	+	-	52	50
0	+	+	49	1
40	-	-	50	47
40	+	+	30	0
120	-	-	28	17
120	-	+	68	34
120	+	-	149	4
120	+	+	192	1

[a]Sucrose concentrations were 0.5% or higher

It has been postulated that the following factors in order of importance influence the potential for botulinal growth in bacon (Tompkins, unpublished data).

1. pH decrease during temperature abuse (i.e. sugar level).
2. brine strength
3. residual nitrite concentration
4. ascorbate/isoascorbate level
5. inoculum _C. botulinum_ level
6. temperature abuse
7. phosphate

These factors can exert synergistic effects with pH very influential in determining the effectiveness of the other factors. A brine strength of 8.6 is required to inhibit _C. botulinum_ type A at pH 6.5, while at pH 5.3, a brine strength of only 4.5 is necessary. Other preservative systems that have been suggested for reduced-nitrite bacon (i.e. potassium sorbate) do show some temporary inhibitory activity, but after prolonged temperature abuse, this inhibition is overcome with the formation of botulinal toxin (Tanaka et al., 1980). Starter cultures provide a natural, consistent pH control mechanism that is activated readily with temperature abuse. A significant advantage of this system is the increasing inhibitory activity. The product actually becomes "safer" during periods of prolonged temperature abuse.

CHAPTER 7
Cured Meats

The term "cured meats" generally has referred to whole cuts of meat (i.e. hams, bellies, shoulders, loins) which are "cured" with a solution of salt, nitrite, nitrate, sugars, sodium erythorbate, flavorings, etc. The meat cuts are either pumped with the solution and/or immersed in the "pickling brine" for a specific time interval. After the curing brine has sufficiently penetrated the meat tissue, the individual products are exposed to varying degrees of heat treatment. Characteristic products are also smoked and/or subsequently dried. "Dry-cured" meats are those whole cuts that are cured with a dry mixture being applied to the surface of the raw material. Although many sausage-type products are also "cured" with salt and nitrite, they generally are classified separately by virtue of their respective processes. The primary distinction between "cured meats" and sausages is the size reduction (i.e. grinding, chopping) exhibited with the latter.

Since the beginning of recorded history, salt has been used for its preserving and dehydrating properties. The coloring, flavoring, and preserving action of nitrates and nitrites also has gained importance. This total action of salt and nitrite on meat tissue has been designated as "curing". The curing, or pickling process, has become less important in recent years as a method of meat preservation with the widespread utilization of refrigeration and packaging techniques. The acceptability of cured meats has persisted, but the marketing demands for individual products have changed. The more traditional cured meats were extremely salty and tough, due to the long dehydration process. The preservation of the meat was the primary objective in the process with any flavor development of a secondary importance. Today, most consumers desire a juicy, slightly salty product that is tender and exhibits the unique flavor and aroma of cured meat.

Increasing technology has resulted in rapid curing techniques that achieve a product in a much shorter time with less dehydration and salt content. The curing brines are also kept at low temperatures (40-50F, 4-10C). These rapid processing methods at low temperatures have also retarded the formation of the typical flavors and aromas that were developed during the "aging" of the more traditional products at higher temperatures (60-75F, 15.5-24C). In addition, the meat raw materials to be used for cured meat products are now less controllable by the processor since the industry receives the meat in pre-cut portions. Traditionally, the small processor would select only superior quality cuts from certain type animals for subsequent cured meat processing.

The current emphasis in cured meat research is to control the biochemical processes and the bacterial growth to achieve a rapid flavor development to accompany the rapid curing techniques. As a result, microbial starter cultures are being investigated and utilized to accelerate and stabilize the "maturation process". The use of microorganisms to accelerate the "aging process" with the development of specific flavors has been successful commercially in the dairy industries for many years.

Bacterial cultures with predetermined metabolic activity are added to the curing brine. These normally consist of a micrococci-type organism to effect nitrate/nitrite reduction (i.e. rapid cure) and limited lipolytic activity (i.e. flavor) and a lactic-type culture for acid production (i.e. flavor and stability). These cultures are added directly, after the other curing ingredients. Traditionally, old brines of good quality used for curing contained an abundance of micrococci-type organisms and some lactic-type organisms. The meat is then "seeded" with the culture(s) when the brine is injected into the meat. If the product is not pumped with the brine, or if meat is dry-cured, the microbial activity generally is limited to the meat surface. However, metabolic end-products (i.e. organic acids, amino acids, nitrite) can diffuse into the internal portion.

Most of the developmental work with starter cultures in cured meat processing is being done by European processors (Coretti, 1977). The "starter culture process" for hams differs from the current processing procedures in that the time for brine immersion is reduced and the subsequent smoking temperatures are increased. The higher temperatures accelerate the microbial and enzymatic activities to shorten the maturation time.

Typically, ham portions are cut as usual and immersed in a brine solution of salt (1.21 to 1.24 grams/cm^3) and nitrite for 24-48 hours. The portions are then injected with another brine solution containing the starter culture, salt, nitrite, and sugar. The "pump percent" as calculated based on the raw weight should not exceed 4 to 5%. Good distribution of the brine internally is essential and very fine needles are desirable to avoid puncture marks and pickle pockets.

Pumping brine (Schiffner et al., 1978)

brine of salt and nitrite (1.12 to 1.14 g/cm^3)	10 liters
starter culture (10^9/ml)	500 ml
sugar	1.2 kg

The starter culture is added after the other dry ingredients have been thoroughly dissolved and just prior to inoculation. Many cultures will lose metabolic activity when exposed to the concentrated brine for prolonged periods of time.

After injection, the ham portions are held in the smokehouse at 25 to 28C (77 to 82F) with 70-80% relative humidity for 18 to 36 hours depending on the size and end-product desired. The humidity is not as critical as required with fermented sausages since the hams are relatively large and retain moisture. Smoke density should be high with reduced air movement. This procedure usually results in hams with a pH of 5.1-5.3, good color and stability. The hams can then be either cooked or further dried.

In attempts to modify dry-curing processes, some European researchers have also recommended the use of starter cultures (Puolanne, 1982). Hams are boned and chunked into 2-3 pound portions and pumped with a 98° salometer pickle containing 0.1% nitrite, 0.2% nitrate, 5% glucose, and starter culture. After injection, the hams are held at 36-39F (2.2-3.9C) for 1-2 days with the surfaces rubbed with coarse salt. After washing, the hams are air dried at 68F (20C) for seven days and vacuum packaged. Total weight loss amounts to about 25%, and the final salt level is in the 3-4% range.

In the United States, the use of starter cultures has been investigated in the production of country-style, dry-cured hams (Bartholomew and Blumer, 1977). Concentrated cultures of P. cerevisiae were injected separately into various parts of the ham so as to achieve sufficient distribution. The volume of water diluent for the culture was insignificant compared to the total water present in the ham. The hams were then dry cured with the other ingredients and processed as commercially practiced. Subjective evaluations indicated that the control and inoculated hams were similar in flavor and other quality characteristics, although aged flavor increased as acidity increased with the "cultured" hams. Generally, white film formation, which is positively correlated with increased crumbliness, softness, and tenderness, increased in inoculated hams with increased aging time. Ham film formation results from free amino acid precipitation (mainly tyrosine) on ham surfaces and probably increases with increasing lactic acid production.

The "aged" flavor associated with many traditional, cured meat products can result from a combination of microbial fermentation, proteolysis, and lipolysis. Many dry-cured hams actually increase in pH due to the hydrolysis of amino acids and the exposure of basic groups to a level where any acid produced by lactic bacteria is neutralized. The aged flavor appears to be associated with the level of free amino acids in country-style ham and the pH of the final ham may be above that of the raw tissue.

Generally, the microbial starter cultures employed commercially by European processors of cured meats have been micrococci-type organisms. Their high salt tolerance, proteolytic and lipolytic activities coupled with a reduced fermentative ability make these microorganisms well suited for cured, whole cuts. In addition, their nitrate-nitrite reducing capability enhances cure color and flavor development since most of these products are not heated to any significant extent. A specific strain of "Micrococcus specialis" that is salt tolerant up to 22% is marketed by a German firm. The strain is active down to pH 4.8, develops catalase and nitrite and nitrate reductase, and converts sugars to acids. The manufacturer recommends the culture for all kinds of hams; also for cooked ham if at least 2-3 days are allowed between injecting and cooking. The culture can be used for "both dry and wet curing, especially with injection curing" (Rudolph Mueller technical literature).

Lactic acid-type microorganisms (i.e. lactobacilli, pediococci) can be used effectively for cured meats where a reduced pH is desirable. Lower meat pH will effect a faster drying rate, thus decreasing the process time to achieve a designated yield (Table 46). Additionally, lower pH will enhance cure color development (Table 47). However, the extent of the fermentation must be controlled to avoid an excessively "tangy" product, which is not desirable in these types of products. This usually is accomplished by limiting the added carbohydrate content and/or a subsequent heat treatment. Besides the flavor contribution, a limited, initial fermentation also can inhibit potential spoilage microorganisms (Table 48). This can enable a processor to reduce the curing time by using higher curing temperatures and/or safely to reduce the salt content. Greater product stability can be achieved at a higher moisture content when the pH is lowered (Bartholomew and Blumer, 1980).

Investigators (Pina, 1977) have demonstrated that a mixed starter culture of Pediococcus cerevisiae and Micrococcus sp. is an effective preservative agent in the manufacture of boiled ham. The cultures were inoculated into the meat with the brine, and the ham was allowed to cure several days prior to cooking. In various commercial trials, the starter culture enhanced product stability through an antagonism toward spoilage microorganisms. Color improvement also was noted in some samples.

Table 46. Acidity and yields of fermented, formed hams after 10 days of aging (Terlizzi et al., 1980).

Fermentation period (hr)	Ham pH	Lactic acid %	Product yield (%)
0	6.4	0.00	87.8
9.5	5.9	0.22	86.9
14.5	5.6	0.41	86.4
18	5.4	0.52	85.8
21	5.2	0.69	85.1

Table 47. Color values of formed hams (adapted from Terlizzi et al., 1980).[1]

Product or stage in process	Gardner color value[2]			a/b ratio
	L	a	b	
Non-fermented, aged	53.7±1.2	9.8±0.6	8.9±0.3	1.10
Fermented mix	50.5±3.2	9.7±2.2	8.6±0.7	1.13
Fermented, heated	52.9±1.4	10.9±0.7	7.3±0.5	1.49
Fermented, aged	53.0±1.4	9.9±0.8	7.1±0.4	1.39

[1]Mean ± standard deviation

[2]Greater redness = higher a, lower b values = higher a/b ratios.

Table 48. A comparison of titratable acidity and pH in dry-cured ham muscle (adapted from Bartholomew and Blumer, 1980).

Inoculum[1]	Titratable acidity (%)	pH
Control	---	5.91
L. plantarum	1.21	5.33
P. cerevisiae	0.97	5.65
S. epidermis	0.95	5.63

[1]Average of six muscle samples from three hams. Microorganisms inoculated in log units of 8.43, 8.23, and 5.68 organisms/g of ham respectively.

CHAPTER 8

The Future of Microorganisms in Meat Processing

The utilization of microorganisms in meat processing will continue to expand and diversify. Although meat preservation by more modern procedures (i.e. refrigeration, freezing, packaging) is not a problem in industrialized countries, fermented sausages still have persisted in civilized societies. The manufacturing procedures have not changed drastically, although the modern industry now has better control over the process through an understanding of the scientific principles involved. Dry sausage production in U.S. federally inspected operations increased 2% during 1981 while total sausage production decreased 3% (AMI, 1982). The production forecast through 1987 also predicts an average increase of 2.1% per year for dried and semi-dried sausage while total sausage products are expected to increase only an average of 1.2% (Anonymous, 1982). According to the FAO, the production and consumption of meat and meat products will increase during the next few years. Growth is expected in industrialized and especially in developing countries. Fermented products will certainly play a key role in this market development since these products have good stability without refrigeration and the initial nutritive value is maintained.

Fermentation as a means of meat preservation is becoming more significant with increasing energy costs for refrigeration, freezing, and/or dehydration. Lowering meat pH provides an effective and economic method to enhance product stability while preserving the nutritive and quality characteristics. Acidity in combination with other preservative techniques (i.e. packaging, heat treatment, reduced water activity) is widely used throughout the food industry. Microbial cultures have proven effective as "acidulation agents" for meats since the relatively slow, consistent, and uniform acid release, via metabolism, does not prohibit the extraction and binding of the soluble meat proteins. Attempts to duplicate microbial action with chemical acidulants, added directly to the meat, have been unsuccessful since direct, rapid acidulation prohibits "bind formation", yielding an unacceptable product texture. In addition, the organic acids, primarily lactic acid, produced by the microorganisms are relatively "mild" and acceptable to the palate. Limited success has been achieved through the use of encapsulated acids that can afford a slow release via heat and/or moisture contact. The capsule material is usually a vegetable fat with a low, specific melting point. The chemical acidulant glucono-delta-lactone (GDL), a ring structure, is used extensively since it affords a slow release of gluconic acid when hydrated. However, the ultimate flavor of gluconic acid is not as desirable as lactic acid in many products, and GDL can acidulate the meat prematurely since hydration often can occur rapidly. Control of product temperature can effectively control microbial cultures.

Further research undoubtedly will concentrate on replacing viable starter cultures with enzymes such as lactic acid-producing enzymes, nitrate/nitrite reductases, catalase, peroxidase, etc. that, theoretically, would yield the same result as the whole microbial cells. However, previous studies generally have been unsuccessful since most of these enzymes are endoenzymes, and it is difficult to recover "active" preparations from broken cells. These endoenzymes usually are part of larger, particle bound systems. They function effectively within the cell, but they function far less effectively within the "hostile" meat environment (i.e. salt, low moisture, curing agents, pH). In addition, enzymes do not multiply, and they do not exhibit other beneficial characteristics such as antagonism towards food pathogens and/or produce additional flavor/aroma compounds. More cost-effective research will focus on new control procedures in vivo to minimize any undesirable characteristics of the viable microorganisms and to optimize the development of the desirable reactions.

Microorganisms are desirable as natural food preservatives which have an extensive record of effectiveness and safety in a wide range of food systems. The use of cultures often can preclude the necessity for other chemical preservatives. Commercially-available meat cultures are approved by the respective regulatory agencies (FDA, USDA) in the United States who also encourage the controlled acidulation of fermented foods. In many European countries meat starter cultures are recognized as "technical ingredients" or "processing aids" and do not require specific approvals. As a result, culture use in fermented meats will increase as more traditional processes convert to the use of starters to achieve greater product consistency and insure safety.

In the future, more specialized cultures will be tailored individually to meet specific product formulation and processing requirements. Historically, meat culture development in the United States has focused on achieving rapid and consistent lactic acid development in fermented sausages. Decreasing fermentation time was of primary importance for economic reasons, and the more rapid acid development achieved with cultures did not appear to adversely affect the quality of the final product. Research has yielded starter cultures with a high level of acid-producing-activity, and many sausage manufacturers have achieved the minimum fermentation times that are practical while still accomplishing other processing objectives (i.e. smoke and heat application, drying, mold growth). New cultures will be designed to yield unique flavor attributes, function as more effective preservatives of color and flavor, and/or retard rancidity development and the proliferation of undesirable microorganisms. Consistent acid production will be maintained through the utilization of culture blends, whereby the specialized culture(s) will be combined with a proficient acid producer(s). The emergence of genetic manipulation techniques probably will contribute to a greater degree of control of microbial characteristics and improve production yields. Inactive and/or immobilized cell preparations are also being explored whereby, in one instance lactic acid producing cells are rendered non-viable by irradiation (Lee et al., 1971). The non-viable cells in combination with a fermentable substrate are inoculated into the meat product, and their acid producing capacity functions to lower product pH when the conditions are favorable (i.e. temperature). Other treatments such as irradiation with beta rays, x-rays or ultraviolet rays, or exposure to high temperatures for a short time may also be used to render the cells non-viable. Treatment with hydroxylamine or other chemicals that inactivate nucleic acids are also conceivable.

Starter cultures will also be expanded to more types of meat products where unique flavors and/or greater stability are desirable. More rapid "aging" processes will emerge, as well as the use of cultures in nonfermented meats. In addition to controlled acidulation, microorganisms will be utilized to control undesirable chemicals (i.e. histamines, nitrosamines, pesticides) and food pathogens.

Although ironic, various segments of the meat industry that have achieved greater sanitation and sterility may eventually revert to the historic, more natural control mechanisms to achieve product safety and stability. The use of microbial cultures will certainly play a role in this total preservative system and allow for greater utilization of perishable, raw meat supplies.

Appendix

pH and Measurement

A pH reading indicates the relative acidity or alkalinity of a substance. The term pH is defined as the negative logarithm of the hydrogen ion (H^+) concentration expressed in moles/liter. The pH scale ranges from 0-14, with a neutral solution as pure water at 25C (77F) having a pH 7. In a neutral solution the hydrogen ion and the hydroxyl ion (OH^-) concentrations are equal at 10^{-7} moles/liter. A solution having a pH less than 7 is termed "acidic" while one with a pH greater than 7 is termed "basic" or "alkaline". A change of 1 pH unit indicates a tenfold change in the hydrogen ion concentration. The increments of pH are logarithmic, not linear. A shift from pH 6 to 5 represents 90 units of hydrogen ion (acid) activity, but to go from 5 to 4 represents 900 units. Midpoint on the pH scale is 0.3 units, not 0.5.

The pH definitely affects meat characteristics and stability. Live animals normally have a pH value between 7.0 - 7.2. After slaughtering, the meat pH decreases due to enzymatic breakdown of glycogen to lactic acid (i.e. glycolysis). The rate of the pH decline depends on several factors including species, genetic characteristics, and stress prior and during slaughter.

Higher pH values (>5.6) indicate the meat tissue will hold more water, however it will be more susceptible to microbial growth. Lower pH values (i.e. greater acidity) indicate reduced water holding capacity, better cure-color development, an acid or "tangy" flavor and greater inhibition against microbial growth, especially meat pathogens.

In the manufacture of fermented products, the meat is acidified deliberately by the use of desirable microorganisms, reducing the pH from 5.6 - 6.2 to 4.7 - 5.3. The lactic acid generated through fermentation provides product stability, flavor, and it allows the meat tissue to release moisture more readily and uniformly since the lower pH is closer to the isoelectric point of meat protein. Most fermented meats are also dried to varying degrees during processing. The lactic acid also denatures meat protein resulting in a "firmer" texture.

The meat pH may be measured by colormetric methods or by pH meters. Due to ever increasing technology, pH meters are readily available that are compact, convenient, and accurate. Paper strips (i.e. pH paper) is another method of measurement, but generally this method is not accurate enough for the pH measurement of meat.

The "slurry method" is the most accurate means of pH measurement of meat products.

1. Standardize pH meter with two buffers (i.e. pH 4 and 7) prior to use. The buffers should be the same temperature as the sample. Leave the probe in distilled water while not in use, and wash the probe with distilled water between each standardization and each measurement.

2. Slice a cross-section of the sausage, or other meat sample as to obtain a uniform sample containing both surface and internal portions (i.e. 50 - 100g).

3. Dice the sample into small pieces with a clean knife. Leave the casing in the sample.

4. Place approximately 50g of sausage in a blender jar, add distilled water (about 150g), and blend at high speed 10 - 15 seconds to make a smooth slurry.

5. Measure pH at the observed temperature of the sample. Many pH meters have automatic temperature compensation.

6. Do not record pH until the probe has stabilized at a constant pH.

7. Remove probe, wash with distilled water, and either read more samples or leave in distilled water. Be sure to wipe off all fat/ tissue from probe and clean periodically with acetone or suitable solvent.

Insertion pH probes that are available for the direct measurement of pH in the samples are effective if correlated with "slurry" measurements. The probes must be also standardized prior to use, well maintained, and completely surrounded by an aqueous medium to provide accurate readings. The continuous measurement of pH during fermentation is also possible with the proper equipment.

NOTE: pH readings are affected by product temperature. Accurate readings can be obtained only when temperature is considered in the measurement.

Bibliography

Acton, J.C. (1977). The chemistry of dry sausages. In Proc. 30th Ann. Recip. Meat Conf. Am. Meat Sci. Assoc., Auburn, Alabama.

Acton, J.C. (1978). Fermented sausages and fermented semi-dry sausages. Chemical and physical properties of the meat. In Proc. 20th Ann. Meat Sci. Inst., Athens, Ga., February 26-March 1.

Acton, J.C. and Dick, R.L. (1975). Improved characteristics for dry fermented turkey sausage. Food Product. Devel. 9:91-94.

Acton, J.C., Dick, R.L. and Norris, E.L. (1977). Utilization of various carbohydrates in fermented sausage. J. Food Sci. 42:174.

AMI (1979). "Meat Facts: A Statistical Summary about America's Largest Food Industry". Am. Meat Inst., Washington, DC.

AMI (1982a). "Meat Facts: A Statistical Summary about America's Largest Food Industry". Am. Meat Inst., Washington, DC.

AMI (1982b). "Good Manufacturing Practices, Fermented Dry and Semi-Dry Sausage". Am. Meat Inst., Washington, DC.

AMI (1982). Newsletter. Am. Meat Inst., Washington, DC. October 15.

Andres, C. (1977). Starter culture for sausage has two microorganisms for better performance. Food Processing. 38(1):132.

Anonymous (1978). "Bactoferment 61, Duploferment 66, Technical Bulletin". Rudolf Müller and Co. Federal Republic of Germany.

Anonymous (1978). Die Fleischwirtschaft. 5:748.

Anonymous (1978). Fast aging with mold spray. Meat Ind., August, p. 51.

Bacus, J. (1978). Bacon processing to achieve reduction in subsequent nitrosamine formation. ABC Research. Technical Report submitted to USDA. October 23.

Bacus, J. (1979). Reduces nitrosamines. Food Engineering. 51(5):24.

Bacus, J.N. and Brown, W.L. (1981). Use of microbial cultures: meat products. Food Technol. January, p. 74.

Bacus, J.N. (1983). Custom Cultures™ Technical Binder. ABC Research, Gainesville, Florida.

Bailey, M.E. and Swain, J.W. (1973). Influence of nitrite on meat flavor. In Proc. Meat Ind. Res. Conf. Am. Meat Inst., Washington, DC.

Baird-Parker, A.C. and Freame, B. (1967). Combined effect of water activity, pH and temperature on the growth of C. botulinum from spore and vegetative cell inocula. J. Appl. Bacteriol. 30:420.

Barber, L.E. and Deibel, R.H. (1972). The effect of pH and oxygen tension on staphylococcal growth and enterotoxin production in fermented sausage. Appl. Microbiol. 24:891-896.

Bartholomew, D.R. and Blumer, T.N. (1977). The use of a commercial Pediococcus cerevisae starter culture in the production of country-style hams. J. Food Sci. 42(2):494.

Bartholomew, D.R. and Blumer, T.N. (1980). Inhibition of S. aureus by lactic acid bacteria in country-style hams. J. Food Sci. 45:420.

Bowen, V.G. and Deibel, R.H. (1974). Effects of nitrite and ascorbate on botulinal toxin formation in wieners and bacon. In Proc. Meat Ind. Res. Conf., Am. Meat Inst. Found., Arlington, VA, p. 63.

Brankova, R. (1976). Antagonism of starter cultures towards some microbial species. In Proc. European Meeting Meat Res. Workers. Sofia, Bulgaria, No. 22, 67:1.

Buchanan, R.E. and Gibbons, N.E. (1974). Bergeys Manual of Determinative Bacteriology. The Williams and Wilkins Company, Baltimore, MD.

Buchanan, R.L. and Soldberg (1972). Interaction of sodium nitrite, oxygen and pH on growth of Staphylococcus aureus. J. Food Sci. 37:81.

Cantoni, A., Bianchi, M.A. and Beretta, G. (1974). Amino acids, histamine and tyramine variation during ripening of dry sausage (salami). Ind. Aliment. 13:75.

Cantoni, C., Molnar, M.R., Renon, P. and Giolitti, G. (1967). Lipolytic micrococci in pork fat. J. Appl. Bacteriol. 30:190.

Caserio, G., Stecchini, M., Pastore, M. and Gennari, M. (1980). The effect of nisin and nitrite, individually and in combination on the germination of spores of Clostridium perfringens in cooked meat products. Institute for the Inspection of Food Stuffs of Animal Origin. U. of Milan, Milan, Italy (unpublished).

CAST (1978). Nitrite in meat curing: risks and benefits. Council for Agricultural Science and Technology. Rep. No. 74, March 6.

Chaiet, L. (1960). Method of fermenting meat products and composition therefor. U.S. Patent 2,945,766.

Checko, P.J., Lewis, J.N., Altman, R., Black, K., Rosenfeld, H. and Parkin, W. (1976). Salmonella bovis-morbificans in precooked roasts of beef. Morbid. Mortal. Weekly Rep. 25(42):333.

Checko, P.J., Lewis, J.N., Altman, R., Halpin, G., Inglis, R., Pierce, M., Pilot, K., Prince, J., Rednor, W., Fleissner, M., Lyman, D., and Parkin, W. (1977). Multi-state outbreak of Salmonella newport transmitted by precooked roasts of beef. Morbid. Mortal. Weekly Rep. 26(34):277.

Christiansen, L.N., Johnston, R.W., Kautter, D.A., Howard, J.W. and Aunan, W.J. (1973). Effect of nitrite on toxin production by C. botulinum and on nitrosamine formation in perishable canned comminuted cured meat. Appl. Microbiol. 25:357.

Christiansen, L.N., Tompkin, R.B., Shaparis, A.B., Kueper, T.V., Johnson, R.W., Kautter, D.A. and Kolari, O.J. (1974). Effect of sodium nitrite on toxin production by Clostridium botulinum in bacon. Appl. Microbiol. 27:733.

Christiansen, L.N., Tompkin, R.B., Shaparis, A.B., Johnston, R.W. and Kautter, D.A. (1975). Effect of sodium nitrite and nitrate on C. botulinum growth and toxin production in summer style sausage. J. Food Sci. 40:488.

Collins-Thompson, D.L., Chang, P.C., Davison, C.M., Larmond, E. and Pivnick, H. (1974). Effect of nitrite and storage temperature on the organoleptic quality and toxinogenesis by C. botulinum in vacuum packaged side bacon. J. Food Sci. 39:607.

Coretti, K. (1977). Starter cultures in the meat industry. Die Fleischwirtschaft 3:386-388.

Daly, C., Sandine, W.E., and Elliker, P.R. (1972). Interactions of food starter cultures and food-borne pathogens. J. Milk Food Technol. 35:349.

Daly, C., Chance, N., Sandine, W.E. and Elliker, P.R. (1973). Control of Staphylococcus aureus in sausage by starter culture and chemical acidulation. J. Food Sci. 38:426.

Deibel, R.H. (1974). Technology of fermented, semi-dried and dried sausages. In Proc. Meat Ind. Research Conf., Am. Meat Inst. Found., Washington, DC.

Deibel, R.H. and Niven, C.F., Jr. (1957). Pediococcus cerevisiae: a starter culture for summer sausage. Bacteriol. Proc. 14-15.

Deibel, R.H., Niven, C.F. and Wilson, D.D. (1961). Microbiology of meat curing. III. Some microbiological and related technological aspects in the manufacture of fermented sausages. Appl. Microbiol. 9:156-165.

De Ketelaere, A., Demeyer, D., Vanderkerckhove, P. and Vervaeke, J. (1974). Stoichiometry of carbohydrate fermentation during dry sausage ripening. J. Food Sci. 39:297-300.

Demeyer, D., Vanderkerckhove, P., Vermeuler, L. and Moermann, R. (1978). Compounds determining pH in dry sausage. In Proc. Europ. Meeting Meat Res. Workers. Belgrade, Yugoslavia.

Dethmers, A.E., Rock, H., Fazio, T. and Johnston, R.W. (1975). Effect of added sodium nitrite and sodium nitrate on sensory quality and nitrosamine formation in thuringer sausage. J. Food Sci. 40:491.

Donnelly, L.S., Ziegler, G.R. and Acton, J.C. (1982). Effect of liquid smoke on the growth of lactic acid starter cultures used in the manufacture of fermented sausage. J. Food Sci. 47:2074-2076.

Drake, E.T. (1928). Process of curing meats. U.S. Patent 1,685,630.

Dryden, F. and Birdsall, J. (1980). Why nitrite does not impart color. Food Technol. 37:7-29.

Dubois, G., Beaumier, H. and Charbonneau, R. (1979). Inhibition of bacteria isolated from ground meat by Streptococcaceae and Lactobacillaceae. J. Food Sci. 44(6):1649.

Eilberg, B.L. and Liepe, H. (1977). Possible improvements in dry sausage technology by adding streptomycetes as starter culture. Die Fleischwirtschaft 9:1698.

Eitenmiller, R.R., Koehler, P.E. and Reagen, J.O. (1978). Tyramine in fermented sausages: factors affecting formation of tyramine and tyrosine decarboxylase. J. Food Sci. 43:699.

Emswiler-Rose, B.S., Johnston, R.W., Harris, M.E. and Lee, W.H. (1980). Rapid detection of staphylococcal thermonuclease on casings of naturally contaminated fermented sausages. Appl. and Environmental Microbiol. 40(1):13.

Everson, C.W., Danner, W.E. and Hammes, P.A. (1970). Bacterial starter cultures in sausage products. J. Agr. Food Chem. 18:570-571.

Everson, C.W., Danner, W.E., and Hammes, P.A. (1974). Process for curing dry and semi-dry sausages. U.S. Patent 3,814,817.

Fazio, T., White R., Dusold, L. and Howard J. (1973). Nitrosopyrrolidine in cooked bacon. J. Assoc. Official Anal. Chem 56:919.

Fiddler, W., Pensabene, J., Fagan, J., Thorton, E., Piotrowski, E. and Wasserman, A.E. (1974). The role of lean and adipose tissue on the formation of nitrosopyrrolidine in fried bacon. J. Food Sci. 39:1070.

Fiddler, W., Pensabene, J., Kushnis, I. and Piotrowski, E. (1973). Effect of frankfurter cure ingredients on n-nitrosodimethylamine formation in a model system. J. Food Sci. 38:714.

Fischer, U. and Schleifer, K.H. (1980). Presence of staphylococci and micrococci in dry sausage. Fleischwirtschaft. 60(5):1046-1048,51.

Ford, J.F. (1980). Alternative bacon processing procedures and their effect on nitrosamine formation. Masters Thesis, U. of Florida.

Fox, J. (1980). Genetic engineering industry emerges. Chemical and Engineering News. 58(11):15.

Fox, J.B. and Nicholas, R.A. (1974). Nitrite in meat. Effect of various compounds on loss of nitrite. J. Agr. Food Chem. 22:302.

Freeman, R.L., Ebert, A.G., Lytle, R.A. and Bacus, J.N. (1982). Effect of sodium nitrite on flavor and TBA values in canned, comminuted ham. J. Food Sci. 47(6):1767.

Genigeorgis, C. (1974). Recent developments on staphylococcal food poisoning. In Discussion of Selected Foodborne Diseases (Edited by C. Genigeorgis). Veterinary Science Division, Academy of Health Science, US Army, San Antonio, TX, 34-92.

Genigeorgis, C.A. (1976). Quality control for fermented meats. J. Vet. Med. Assoc. 169(11):1220.

Genigeorgis, C., Foda, M.S., Mantis, A. and Sadler, W.W. (1971). Effect of sodium chloride and pH on enterotoxin C production. Appl. Microbiol. 21:862.

Gilliland, S.E. (1980). The use of lactobacilli to preserve fresh meat. In Proc. Recip. Meat Conf. Amer. Assoc. of Meat Sci. 33:54.

Gilliland, S.E. and Speck, M.L. (1975). Inhibition of psychrotropic bacteria by lactobacilli and pediococci in non-fermented refrigerated foods. J. Food Sci. 40:903.

Goepfert, J.M. and Chung, K.C. (1970). Behavior of salmonella during the manufacture and storage of a fermented sausage product. J. Milk Food Technol. 33:185.

Goepfert, J.M. (1980). Salmonella in foods - update. Presented at the A.B.C. Research 6th Annual Technical Seminar, Gainesville, FL, February 17-19.

Gray, J. and Dugan, L. (1975a). Inhibition of n-nitrosamine formation in model food systems. J. Food Sci. 40:981.

Greenberg, R.A. (1975). Update on nitrite, nitrate and nitrosamines. Proc. Meat Ind. Res. Conf.:71.

Gryczka, A. and Shah, R.B. (1979). Process for the treatment of meat with compositions including M. varians and a lactic acid producing bacteria. U.S. Patent 4,147,807.

Haines, W.C. and Harmon, L.G. (1973). Effect of lactic acid bacteria on growth of Staphylococcus aureus and production of enterotoxin. Appl. Microbiol. 25:436.

Haldane, J. (1901). The red color of salted meat. J. Hyg. Camb. 1:115.

Herborg, L. and Johansen, S. (1977). Fish cheese: the preservation of minced fish by fermentation. In Proc. Conf. on Handling, Processing and Marketing of Tropical Fish, Ministry of Fisheries, Denmark, p. 253.

Hill, W.M. (1972). The significance of staphylococci in meats. In Proc. 25th Ann. Recip. Meat Conf., National Live Stock Meat Board. Chicago, Il. 300-304.

Hirsch, A. and Wheater, L.M. (1951). The production of antibiotics by staphylococci. J. Dairy Res. 18:193.

Houston, D. (1979). Acid-producing microorganisms in meat products for nitrite dissipation. Fed. Reg. 44(31):9372.

Hurst, A. (1973). Microbial antagonisms in food. Can. Inst. Food Sci. Technol. J. 6:80.

Ingolf, F.N. and Skjelkvale (1982). Effect of natural spices and oleoresins on Lactobacillus plantarum in fermentation of dry sausage. J. Food Sci. 47:1618.

Jay, J.M. (1970). Modern Food Microbiology. Van Nostrand Reinhold Company, New York, NY.

Jensen, L.B. and Paddock, L.S. (1940). Sausage treatment. U.S. Patent 2,225,783.

Johnston, R.W. (1980). Coagulase positive staphylococci update. Presented at the A.B.C. Research 6th Annual Technical Seminar, Gainesville, Fl., February 17-19.

Jonston, M.A., Pivnick, H. and Samson, J.M. (1969). Inhibition of Clostridium botulinum by sodium nitrite in a bacteriological medium and in meat. Can. Inst. Food Technol. J. 2:52.

Jordan, D., Martini, R., McClosky, D., Suroweic, J., Caparelli, P., Grun, D., Altman, R., Coleman, C., Dennis, R., Goldfield, M., Pilot, J., Rednor, W., Rosenfeld, H., Stemhagen, A., Sussman, O. and Timko, F. (1976). Salmonella saint-paul in pre-cooked roasts of beef. Morbid. Mortal. Weekly Rep. 25:34.

Kiernat, B.H., Johnson, J. and Siedler, A. (1964). A summary of the nutrient content of meat. Bulletin No. 57. Am. Meat Inst. Found., Chicago, Il.

Klettner, P.G. (1980). Firmness changes during raw sausage aging. Die Fleischerei. October.

Kurk, F.W. (1921). Art of curing meat. U.S. Patent 1,380,068.

Lawrence, R.C. and Thomas, T.D. (1979). The Fermentation of Milk by Lactic Acid Bacteria. In Microbial Technology: Current State, Future Prospects. Vol. 29, p. 190. Cambridge Univ. Press, New York.

Lechowich, R.V. (1978). In The Science of Meat and Meat Products (Edited by J.F. Price and B.S. Schweigert). p. 262, Food and Nutrition Press, Westport, Conn.

Lee, M. and Cassens, R.G. (1980). Effect of sodium chloride on residual nitrite. J. Food Sci. 45:879.

Lee, W.H., Reimann, H.P. and Al-Mashat, A.J. (1971). Controlled fermentation and prevention of undesirable bacterial growth in food. U.S. Patent 3,794,739.

Leistner, L. (1972). Starterkulturen and schimmelpilze. Niederschrift Starterkultur-Symposium, Helsinki.

Leistner, L. and Rodel, W. (1975). The significance of water activity for microorganisms in meats. In Water Relations of Foods (Edited by R.B. Duckworth), p. 309. Academic Press, New York.
Liebetrau, B. and Grossmann, G. (1976). The role of starter cultures used for raw sausage ripening in the context of food hygiene. Hygienist. Res. Beyirkes Suhl/Sitz Gotha. German Democratic Republic. 20(5):489.
Magee, P.N. and Barnes, J.M. (1956). The production of malignant primary hepatic tumors in the rat by feeding dimethylnitrosamine. Brit. J. Cancer. 10:114.
Masters, B.A. (1979). Fate of Salmonella inoculated into fermented sausage. M.S. thesis, U. of Florida, Gainesville, Fl.
Mayer, L., Freeman, B., Watson, J.C. and Carraway, C. (1975). Salmonella singapore - New Orleans. Morbid. Mortal. Weekly Rep. 24(47):397.
Mihalyi, V. and Körmendy, L. (1967). Changes in protein solubility and associated properties during the ripening of Hungarian dry sausages. Food Technol. 27:1398.
Minor, T.E. and Marth, E.H. (1972). Staphylococcus aureus and staphylococcal food intoxications. A review III. Staphylococci in dairy foods. J. Milk and Food Technol. 35(2).
Minor, T.E. and Marth, E.H. (1972). Staphylococcus aureus and staphylococcal food intoxications. A review IV. Staphylococci in meat, bakery products, and other foods. J. Milk and Food Technol. 35(4).
Mirvish, S. (1970). Kinetics of dimethylamine nitrosation in relation to nitrosamine carcinogenesis. J. Nat. Cancer Inst. 44:633.
Moon, N.J., Beuchat, L.R. and Hays, E.R. (1980). Evaluation of lactic acid bacteria for extending the shelf-life of shrimp. Presented at 40th Ann. Meeting, Inst. of Food Technolog., New Orleans.
Mottram, D. and Rhodes, D. (1973). Nitrite and the flavor of cured meat. Proc. Int. Symp. Nitrite Meat Prod.:161.
Mottram, D., Patterson, R., Edwards, R. and Gough, T. (1977). The preferential formation of volatile n-nitrosamines in the fat of fried bacon. J. Sci. Food Agr. 28:1025.
Nakagawa, A. and Kitahara, K. (1959). Taxonomic studies of the genus Pediococcus. J. Gen. Appl. Microbiol. 5:95-126.
NAS (1975). Prevention of microbial and parasitic hazards associated with processed foods. Ch. 7 Fermented Foods, Fermented Sausages. Natl. Acad. of Sciences, Washington, DC.
National Center for Disease Control (1971). Gastrcenteritis associated with salami. Morbid. Mortal. Weekly Rep. 20:253-258.
National Center for Disease Control (1971). Gastroenteritis associated with Genoa salami. Morbid. Mortal. Weekly Rep. 20:261-266.
National Center for Disease Control (1971). Gastroenteritis attributed to Hormel San Remo stick Genoa salami. Morbid. Mortal. Weekly Rep. 20:370.
National Center for Disease Control (1972). Staphylococcal food poisoning. Morbid. Mortal. Weekly Rep. 21:169-170.

National Center for Disease Control (1975). Salmonella newport contamination of hamburger - United States. Morbid. Mortal. Weekly Rep. 24:438.

National Center for Disease Control (1979). Staphylococcal food poisoning associated with Genoa and Hard Salami - United States. Morbid. Mortal. Weekly Rep. 28(15):179.

Niinivaara. F.P. (1955). The influence of pure bacterial cultures on aging and changes of the red color of dry sausage. Suomen Maataloustieteellisen Seuran Julkaisuja No. 84, Acta Agralis Fennica (Helsinki).

Niinivaara, F.P. and Pohja, M.S. (1954). Zur theorie der wasserbindung des fleisches, Fleischwirtschaft 6, 192.

Niinivaara, F.P., Pohja, M.S. and Komulainen, S.E. (1964). Some aspects about using bacterial pure cultures in the manufacture of fermented sausages. Food Technol., February, p. 25.

Niskanen, A. and Nurmi, E. (1976). Effect of starter culture on staphylococcal enterotoxin and thermonuclease production in dry sausage. Appl. Microbiol. 34(1):11.

Niven, C.F. (1961). Microbiology of meats. Cir. No. 68, Am. Meat Inst. Found., Washington, DC.

Niven, C.F. Jr., Deibel, R.H. and Wilson, G.D. (1958). The AMIF sausage starter culture. Cir. No. 41, Am. Meat Inst. Found., Chicago, IL.

Niven, C.F. Jr., Deibel, R.H. and Wilson, G.D. (1959). Production of fermented sausage. U.S. Patent 2,907,661.

Nordal, J. and Slinde, E. (1980). Characteristics of some lactic acid bacteria used as starter cultures in dry sausage production. Appl. and Environ. Microbiol. 40(3):472-475.

Nurmi, E. (1966). Effect of bacterial inoculations on characteristics and microbial flora of dry sausage. Acta Agr. Fenn. 108:1.

Oxford, A.C. (1944). Diplococcin, an antibacterial protein elaborated from certain milk streptococci. Biochem. J. 38:178.

Park, H.S. and Marth, E.H. (1972). Behavior of Salmonella typhimurium in skim milk during fermentation by lactic acid bacteria. J. Milk Food Technol. 35:482.

Pederson, C.S. (1949). The genus Pediococcus. Bacteriol. Revs. 13:225-232.

Pederson, C.S. (1979). Microbiology of Food Fermentations (2nd ed.). AVI Publishing Co., Westport, Conn.

Peterson, A.C., Black, J.J. and Gunderson, M.F. (1962). Staphylococci in competition. II. Effect of total numbers and proportion of staphylococci in mixed cultures on growth in artificial medium. Appl. Microbiol. 10:23.

Peterson, A.C., Black, J.J. and Gunderson, M.F. (1964). Staphylococci in competition. III. Influence of pH and salt on staphylococcal growth in mixed populations. Appl. Microbiol. 12:70.

Peterson, A.C., Black, J.J. and Gunderson, M.J. (1962). Staphylococci in mixed cultures on growth in artificial medium. Appl. Microbiol. 10:23.

Pelczar, M.J. and Reid, R.D. (1958). Ch. 10. Chemical changes produced by bacteria. Microbiology. The McGraw-Hill Book Co., Inc., New York.

Pensabene, J., Fiddler, W., Gate, R., Fagan, J. and Wasserman, A.E. (1974). Effect of frying and other cooking conditions on nitrosopyrrolidine formation in bacon. J. Food Sci. 39:314.

Petaja (1977). Starter cultures in frankfurter type sausage. Die Fleischwirtschaft 57:109.

Pezold, H.V. (1969). Verderben und vorrathaltung von fette und fettprodukte, in Handbuch der Lebensmittelchemie, Vol. IV (Edited by J. Schormüller), Springer Verlag, Berlin.

Pfeil, E. and Liepe, H. (1973). Can enzymes be obtained from bacteria used for maturing dry sausage? Die Fleischwirtschaft 53:221.

Pina, V.J.G. (1977). Bacterial antagonism--the use in the preservation of boiled meat products. Die Fleischerei. 28(10):43.

Porubcan, R.S. and Sellars, R.L. (1979). In Microbial Technology, Vol. 1, Second Edition (Edited by H.J. Peppler and D. Perlman), p. 59, Academic Press, New York.

Price, J.F. and Schweigert, B.S. (1978). Ch. 4 Muscle Function and post-mortem changes. The Science of Meat and Meat Products. Food and Nutrition Press, Westport, CT.

Puolanne, E. (1982). Dry-cured hams--European style. 35th Ann. Recip. Meat Conf., Amer. Meat Sci. Assoc., Blacksburg, VA, p. 49.

Raccach, M. (1981). Control of Staphylococcus aureus in dry sausage by a newly developed meat starter culture and phenolic-type antioxidants. J. Food Protect. 44:665-669.

Raccach, M. (1981). Method and bacterial compositions for fermenting meats. U.S. Patent 4,303,679.

Raccach, M. and Baker, R.C. (1978). Formation of hydrogen peroxide by meat starter cultures. J. Food Protect. 41:798.

Raccach, M. and Baker, R.C. (1978). Lactic acid bacteria as an antispoilage and safety factor in cooked, mechanically deboned poultry meat. J. Food Protect. 41(9):703.

Reddy, S.G., Hendrickson, R.L. and Olsen, H.C. (1970). The influence of lactic cultures on ground beef quality. J. Food Sci. 35:787.

Rheinbaben, K.V. and Hadlok, R. (1979). Differentiation of microorganisms of the family Micrococcaceae isolated from dry sausages. Die Fleischwirtschaft 59(9):1321.

Rhia, W.E. and Solberg, M. (1975). Clostridium perfringens inhibition by sodium nitrite as a function of pH, inoculum size and heat. J. Food Sci. 40:439.

Rice, S., Eitenmiller, R.R. and Koehler, P.E. (1975). Histamine and tyramine content of meat products. J. Milk Food Technol. 38:256.

Rice, S.L. and Koehler, P.E. (1976). Tyrosine and histidine decarboxylase activities of Pediococcus cerevisiae and Lactobacillus species and the production of tyramine in fermented sausages. J. Milk Food Technol. 39:166.

Riemann, H. (1969). Food-borne Infections and Intoxications. Academic Press, New York, NY.

Riemann, H.P., Lee, W.H. and Genigeorgis, C. (1972). Control of Clostridium botulinum and Staphylococcus aureus in semi-preserved meat products. J. Milk Food Technol. 35:514-523.

Roberts, T.A. and Ingram, M. (1973). Inhibition of growth of C. botulinum at different pH values by sodium chloride and sodium nitrite. J. Food Technol. 8:467.

Roberts, T.A. and Smart, J.L. (1974). Inhibition of Clostridium spp. by sodium nitrite. J. Appl. Bact. 37:261.

Roberts, T.A. and Smart, J.L. (1976). The occurrence and growth of Clostridium spp. in vacuum-packed bacon with particular reference to Cl. perfringens (Welchii) and Cl. botulinum. J. Food Technol. 12:58-62.

Rothchild, H. and Olsen, R.H. (1971). Process for making sausage. U.S. Patent 3,561,971.

Rowland, J.R. and Grasso, P. (1975). Bacterial degradation of nitrosamines. Biochem. Soc. Trans. 3(1):185.

Rust, R.E. (1977). Sausage and Processed Meats Manufacturing. Am. Meat Inst., Washington, DC.

Salzer, U.J., Broeker, U., Klie, H.F. and Liepe, H.U. (1977). Effect of pepper and pepper constituents on the microflora of sausage products. Fleischwirtschaft 57(11):2011-2014.

Scalan, R.A. (1975). N-nitrosamines in foods. Crit. Rev. Food Tech. 5:357.

Schiffner, E., Hagedorn, W. and Oppel, K. (1978). Bakterienkulturen in der Fleischindustrie. VEB Fachbuchvelag, Leipzig.

Schneider, H. (1980). Dry and semi-dry sausage technology primer. Meat Industry, October, p. 62.

Schubring, R. and Kuhlmann, W. (1978). Preliminary studies on application of starter cultures in the manufacture of fish products. Lebensmittel-Industrie 25(10):455.

Schut, J. (1978). The European sausage industry. In Proc. 31st Ann. Recip. Meat Conference. Amer. Assoc. Meat Sci., U. of Conn.

Sen, N., Donaldson, B., Eyengar, J. and Panalaks, T. (1973). Nitrosopyrrolidine and dimethylnitrosamine in bacon. Nature. 241:473.

Sen, N., Iyengar, J., Donaldson, B. and Panalaks, T. (1974). Effect of sodium nitrite concentration on the formation of nitrosopyrrolidine and dimethylnitrosamine in fried bacon. J. Agr. Food Chem. 22:540.

Sen, N., Seaman, S. and Miles, W.F. (1979). Volatile nitrosamines in various cured meat products: Effect of cooking and recent trends. J. Agr. Food Chem. 27:1354.

Sirviö, P. and Nurmi, E. (1977). The effect of starter cultures and various additives on the growth of Salmonella seftenberg in dry sausage. Die Fleischwirtschaft 5:1007.

Smith, J.L., Huhtanen, C.N., Kissinger, J.C. and Palumbo, S.A. (1975). Survival of salmonellae during pepperoni manufacture. Appl. Microbiol. 30:759-763.

Smith, J.L., Palumbo, S.A., Kissinger, J.C. and Huhtanen, C.N. (1975). Survival of Salmonella dublin and Salmonella typhimurium in Lebanon bologna. J. Milk Food Technol. 38:150.

Smith, J.L. and Palumbo, S.A. (1981). Microorganisms as food additives. J. Food Protect. 44:936-955.

Sofos, J.N., Busta, F. and Allen, C. (1979). Botulism control by nitrite and sorbate in cured meats. A review. J. Food Prot. 42:739.

Stolić, D.D. (1975). Quantitative relationship between micrococci and lactobacilli during ripening of fermented sausages and factors influencing this relationship. Acta Vet Beograd. 25:91.

Storrs, A.B. (1980). Non-frozen concentrated bacterial cultures. U.S. Patent 8,226,940.

Surkiewicz, B.F., Johnston, R.W., Elliott, R.P. and Simmons, E.R. (1972). Bacteriological survey of fresh pork sausage produced at establishments under federal inspection. Appl. Microbiol. 23: 515-520.

Surkiewicz, B.F., Harris, M.E., Elliott, R.P., Macaluso, J.F. and Strain, M.M. (1975). Bacteriological survey of raw beef patties produced at establishments under federal inspection. Appl. Microbiol. 29:331-334.

Sutic, M. (1978). The effect of additives on pure and conjoint cultures for sausages. In Proc. Europ. Meeting Meat Res. Workers, Belgrade, Yugoslavia.

Swift, C.E. and Ellis, R. (1956). The action of phosphates in sausage products. I. Factors affecting the water retention of phosphate-treated ground meat. Food Technol. 10:546.

Tanaka, N., Traisman, E., Lee, M.H., Cassens, R.G. and Foster, E.M. (1980). Inhibition of botulinum toxin formation in bacon by acid development. J. Food Protect. 43(6):450.

Tannenbaum, S. and Fan, T. (1973). Uncertainties about nitrosamine formation in and from foods. Proc. Meat Ind. Res. Conf.:1.

Tatini, S.R., Lee, R.Y., McCall, W.A. and Hill, W.M. (1976). Growth of Staphylococcus aureus and production of enterotoxins in pepperoni. J. Food Sci. 41:223.

Taylor, S.L., Leatherwood, M. and Lieber, E.R. (1978). A survey of histamine levels in sausages. J. Food Protection. 41(8):634.

Terlizzi, F.M., Acton, J.C. and Skelley, G.C. (1980). Comparison of non-fermented and fermented hams formed with muscle strips, cubes and ground meat. Tech. Contribution No. 1722, South Carolina Ag. Exp. Station, Clemson.

Terrell, R.N., Smith, G.C. and Carpenter, Z.L. (1978). Practical manufacturing technology for dry and semi-dry sausage. In Proc. 20th Ann. Meat Sci. Inst., Athens, Ga.

Tezcan, I. and Yuecel, A. (1975). Effect of a Lactobacillus starter culture in vacuum packaged meat on Salmonella and Pseudomonas organisms. Veteriner Hekimler Dernegi Dergisi 45(4):7-10.

Troller, J.A. and Frazier, W.C. (1963). Repression of Staphylococcus aureus by food bacteria. Appl. Microbiol. 11:163.

UN (1979). Workshop--Research and Development Needs in the Field of Fermented Foods. International Symposium of Microbiological Aspects of Food Storage, Processing and Fermentation in Tropical Asia. Bogor, Indonesia, December 14-15.

United States Department of Agriculture (1973). Meat and Poultry Inspection Regulations. U.S. Government Printing Office, Washington, DC.

Urbaniak, L. and Pezacki, W. (1975). Die Milschsäure bildende Rohwurst-Mikroflora und ihre technologisch bedingte Veranderung. Fleischwissensch 55:229.

USDA (1977). The staphylococcal enterotoxin problem in fermented sausage. Task Force Report. F.S.Q.S., Washington, DC.

USDA (1980). Study to survey nitrosamine levels in dry cured bacon, hams, and shoulders. F.S.Q.S., Washington, DC, February 11.

Zaika, L.L. and Kissinger, J.C. (1979). Effects of some spices on acid production by starter cultures. J. Food Protect. 42(7): 572-576.

Zaika, L.L. and Kissinger, J.C. (1982). Fermentation enhancement by spices: identification of active component. In Proc. 42nd Ann. IFT Meeting, Las Vegas.

Zaika, L.L., Zell, T.E., Palombo, S.A. and Smith, J.L. (1978). Effect of spices and salt on fermentation of Lebanon bologna-type sausage. J. Food Sci. 43(1):186-189.

Zimmerman, W.J. (1970). Prevalence of *Trichinella spiralis* in commercial pork sausage. Public Health Rep. 85:717-724.

Zimmerman, W.J. and Zinter, D.E. (1971). The prevalence of trichiniasis in swine in the United States 1966-1970. Public Health Rep. 86:237-945.

Zottola, E.A. (1972). *Introduction to Meat Microbiology*, AMI Center for Continuing Education, Am. Meat Inst., Washington, DC.

Glossary

Fermented meat processors have developed specialized terms that refer to their unique products, processes, and problems. To communicate effectively, the respective terminology must be understood, and a partial listing of these terms is provided below:

1. acclimate, acclimation - the time interval required for (microbial) adjustment to a new environment prior to effecting the desired reactions.
2. active (culture) - refers to the microbial activity, or relative ability to effect the desired result.
3. aged, aging - refers to the holding of meat at temperatures that allow microbial/chemical reactions to proceed.
4. attack (microbial) - the enzymatic hydrolysis of protein or fat in the meat or sausage casing (i.e. natural, collagen).
5. backinoculum, backslop - the material/process of inoculating a new batch with a previously fermented batch that contains high numbers of microorganisms.
6. batter - the formulated raw sausage mix, prior to any heat processing.
7. bind - the ability of the meat proteins to hold the product intact by emulsifying fat and binding water.
8. breaking (of meat) - the size reduction of meat units to afford uniform mixing, formulation.
9. buffering capacity - the capacity of the meat mix to absorb H^+ ions (i.e. acid) prior to demonstrating a decreased pH.
10. bung - straight large intestine of the hog, etc.
11. carrier - refers to a material (i.e. salt, dextrose) used as a diluent that affords better distribution of a minute amount of a concentrated ingredient.
12. case-hardening - the condition in a meat product whereby the surface tissue has exhibited a more rapid moisture loss than the internal portion, resulting in the formation of a tough "crust" that retards subsequent moisture loss.
13. chance inoculation - the random introduction into a product of microorganisms that are present in the specific environment.
14. charring - the Mallaird reaction, whereby the amino groups of proteins react with reducing sugars, generally at high temperatures, to yield a browning discoloration in meat products.

15. collapsing - a condition exhibited in dry sausages where "case hardening" has occurred. The internal product moisture subsequently escapes rapidly and non-uniformly with the "collapse" of the rigid product surface--leaving irregular surface grooves.
16. come-up-time - the time required for the internal product temperature to reach the external processing temperature.
17. controlled-atmosphere - the processing environment is controlled as to temperature, relative humidity, and air circulation.
18. cultured - a product and/or process utilizing a prepared starter culture.
19. cure, curing - the ingredients or process of adding salt, nitrate, nitrite, sugar and/or other ingredients to meat for the purpose of preserving and flavoring.
20. drip room - a chamber with high humidity where dry sausages are fermented in the initial part of the processing sequence. As the cold meat is heated, the product "drips" moisture.
21. drip loss - the weight loss (i.e. moisture) that the sausage exhibits during fermentation.
22. dry curing - the process where dry ingredients are added directly to the meat for the purpose of curing.
23. dry room - a chamber where meat products are held to lose moisture, generally under controlled atmospheric conditions.
24. dry sausage - a sausage, typically losing 30-50% of its initial weight during processing.
25. finish temperature - the maximum internal product temperature achieved during processing.
26. greasing - the loss of fat during processing; an undesirable condition in fermented sausage.
27. green room - a chamber with high humidity where sausages are fermented in the early part of the process (i.e. drip room).
28. lactics - a general term referring to microorganisms producing lactic acid from the fermentation of glucose (i.e. lactic acid microorganisms).
29. lag time - the time required prior to observing a pH decrease in fermented meat.
30. lay down - a processing stage whereby the formulated meat mix is held below 50F (10C) prior to stuffing to enhance cure color development, binding, and development of a lactic microflora.
31. load - the amount of product processed at one time.
32. matched bellies, etc. - two primal cuts from the same animal.

33. maturing room - the chamber for holding meat (i.e. process) under conditions that favor the desired chemical and microbiological reactions.
34. mixed cure - containing both nitrite and nitrate.
35. mother batch - the source of the microorganisms for a back inoculum; a previously fermented batch.
36. paired bellies, etc. - two primal cuts from same animal (i.e. matched).
37. pan curing - traditional process of curing meat in shallow pans to enhance nitrate reduction by micrococci-type microorganisms.
38. pickle - a curing brine; a solution of salt and other ingredients.
39. pit house - a smoking/heating chamber for meat with a "pit" below a grid floor for the burning of sawdust.
40. pores (meat tissue) - generally refers to surface tissue and the ability to transmit moisture.
41. positioning - the arrangement of meat products in a processing chamber.
42. pre-conditioning - the process of treating raw meat prior to the formulation with other ingredients (i.e. temperature adjustment, moisture adjustment).
43. pump - the injection of a curing solution into meat tissue.
44. relative humidity - quantity of water vapor present in the atmosphere to the quantity which would saturate at the existing temperature.
45. rework - final or partially processed product "added back" into the raw batch.
46. ripening - the process of holding a meat product as to effect the desirable chemical and microbial reactions (i.e. aging, maturing).
47. salting - the process of adding sodium chloride to raw meat tissue to preserve and flavor the product.
48. seeding - the inoculation of microorganisms into a raw batch via a back-slop technique.
49. semi-dry sausage - a fermented sausage, typically losing 10-20% of its initial weight during processing.
50. smearing (of fat) - the loss of fat-particle definition whereby the fat tissue encases the lean tissue.
51. soft product - an undesirable condition in dry sausage where the internal portion fails to dry properly, usually due to "case hardening".

52. stability (shelf-life) - the ability of a meat product to maintain its initial characteristics during subsequent storage.

53. stuff (ed) - the insertion of a raw sausage mixture into the casing.

54. surface skin - surface tissue dehydration and/or protein coagulation forming an outer crust.

55. tang - the acidic taste in fermented meats, generally due to lactic acid.

56. traditional process - prior to the use of microbial starter cultures.

57. water activity - mole fraction of the solvent ÷ moles solute + moles solvent (equilibrium relative humidity, ERH, ÷ 100).

58. wild (microorganisms) - those randomly occurring in the environment.

59. yield - the weight of the product after processing divided by the initial weight x 100%.

Index